SpringerBriefs in History of Science and Technology

W0225609

The *SpringerBriefs in the History of Science and Technology* series addresses, in the broadest sense, the history of man's empirical and theoretical understanding of Nature and Technology, and the processes and people involved in acquiring this understanding. The series provides a forum for shorter works that escape the traditional book model. SpringerBriefs are typically between 50 and 125 pages in length (max. ca. 50.000 words); between the limit of a journal review article and a conventional book.

Authored by science and technology historians and scientists across physics, chemistry, biology, medicine, mathematics, astronomy, technology and related disciplines, the volumes will comprise:

1. Accounts of the development of scientific ideas at any pertinent stage in history: from the earliest observations of Babylonian Astronomers, through the abstract and practical advances of Classical Antiquity, the scientific revolution of the Age of Reason, to the fast-moving progress seen in modern R&D;
2. Biographies, full or partial, of key thinkers and science and technology pioneers;
3. Historical documents such as letters, manuscripts, or reports, together with annotation and analysis;
4. Works addressing social aspects of science and technology history (the role of institutes and societies, the interaction of science and politics, historical and political epistemology);
5. Works in the emerging field of computational history.

The series is aimed at a wide audience of academic scientists and historians, but many of the volumes will also appeal to general readers interested in the evolution of scientific ideas, in the relation between science and technology, and in the role technology shaped our world.

All proposals will be considered.

Carlos Sanhueza-Cerda

The Day Laborers of Science. Technical Work at the Astronomical Observatory of Chile (1852–1927)

 Springer

Carlos Sanhueza-Cerda
Universidad de Chile
Santiago, Chile

ISSN 2211-4564   ISSN 2211-4572 (electronic)
SpringerBriefs in History of Science and Technology
ISBN 978-3-031-84349-5   ISBN 978-3-031-84350-1 (eBook)
https://doi.org/10.1007/978-3-031-84350-1

This Springer imprint is published by the registered company Springer Nature Switzerland AG
The registered company address is: Gewerbestrasse 11, 6330 Cham, Switzerland

If disposing of this product, please recycle the paper.

To Ximena, the kaleidoscope of my days

Acknowledgments

I am deeply grateful for the invaluable contributions and unwavering support of Lorena Valderrama, Verónica Ramírez, José Soto, and Stefan Meier. Their dedication, expertise, and friendship have been instrumental in bringing this book to fruition. I am forever indebted to them for their invaluable assistance and encouragement throughout this journey.

I am also grateful for the bibliographical support of David Edgerton, Sergio Grez, and Guillermo Guajardo.

This book was made possible by FONDECYT Grant 1170625.

About This Book

The proposed book explores the materiality of knowledge and the role of labor in astronomy research. The case study centers on Chile's primary astronomical observatory from the late 19th century to the early 20th century, a period during which the observatory was entirely publicly funded. The historiographic aim of the volume is to challenge the conventional narrative of scientific progress, which often emphasizes the contributions of great minds or "sages," thus overlooking the significant role played by the laborers who contribute to scientific endeavors on a daily basis.

This volume straddles the intersection of the history of science and technology, as well as epistemology, providing a comprehensive reflection on how to approach the writing of history in these fields.

Contents

Chapter 1
Introduction

Abstract Here, the problem of the study of astronomy in Chile is installed from the role of those who have surrounded the astronomers: precision mechanics, calculators and observers, as well as builders. This chapter explores the significance of materiality in technical work, challenging the marginalization of material culture in society. It advocates for a shift from idealist notions of knowledge to recognizing the importance of material mediation in disciplinary practices. Emphasizing the role of technicians and collaborative labor reveals hidden aspects of scientific production, questioning traditional narratives of innovation and the heroism of individual scholars. By examining the relationship between technicians and materiality, it reveals the importance of maintenance and repair work in sustaining scientific practice and challenges the dichotomy between intellectual and manual labor in Western history. This perspective highlights the fragility of technological systems and the necessity of constant care and attention to maintain social order.

Keywords Technical knowledge · Materiality · The silenced of history

"There is a profound bias in the literature towards the study of scientists and technologists employed in research, and the distinct silence on employment in other forms of work, such as teaching, routine testing, management, maintenance and so on." David Edgerton, "From Innovation to Use: Ten Eclectic Theses on the Historiography of Technology," in: *History and Technology: An International Journal*, vol. 16, no. 2, p. 125.

This book examines certain threads of the sociomaterial fabric that articulated a set of institutions—global science networks—with instruments, astronomers, technicians and collaborators in the working space of the National Astronomical Observatory of Chile, from its founding in 1852 up until the early decades of the twentieth century. We call this stage "early" as the institution depended on the national government, being under the control of the Ministry of Public Instruction and the supervision of the University of Chile. Once this link was severed in 1927, the National Observatory began to forge a path of its own, very similar to that of the institution we know today. The historiography has understood this period through

© The Author(s), under exclusive license to Springer Nature Switzerland AG 2025 1
C. Sanhueza-Cerda, *The Day Laborers of Science. Technical Work at the Astronomical Observatory of Chile (1852–1927)*, SpringerBriefs in History of Science and Technology, https://doi.org/10.1007/978-3-031-84350-1_1

the role of the astronomers who directed the observatory during the foundational stage of astronomy in Chile. The very chronology has been based around each one of these individuals: Karl Moesta (1852–1867), José Ignacio Vergara (1867–1889), Hubert Obrecht (1889–1906) and Friedrich Ristenpart (1906–1913).[1] We know the astronomers, it's time to peek behind the curtains in order to discover the actions of secondary characters: the technicians, collaborators, computers, builders of buildings and domes.

In January 1912, Richard Wüst, a precision mechanic at the National Astronomical Observatory of Chile, wrote a letter to the newspaper *La Razón* in which he bitterly complained about its coverage of the observatory's work. Wüst responded: "If the author of the article feels that a mechanic at a great observatory is a last-class employee, he only reveals his absolute ignorance of said employee's work." Wüst's dignification of technical work went beyond the specific case of the observatory: "I would like to remind the gentleman who wrote the article in question that Chile's railways waste a great deal of material due to their total lack of trained technical personnel." Wüst then got to the heart of the problem: "Does this gentleman believe (…) that all the employees of a scientific institution are sages? No, sir, they're nothing but the day laborers of science…".[2]

In routine scientific work, everyone is a *day laborer*: technicians, collaborators, auxiliaries and the workers who build and adapt the huts and domes, not to mention the astronomers themselves. Astronomical observations depend on their joint efforts: so that instruments are well-calibrated, so that domes move smoothly and so that dust and mold don't affect the lenses. The stability of an observatory could collapse because a telescope chair was installed incorrectly, if calculations were confused or simply because electric cables had been gnawed by rats. Here, both the *last-class employee* and the *sage* matter. And so why have we only ever seen the latter? There are reasons why the history of astronomy in Chile has focused on astronomers (in part because they are the ones who have written it), but there's almost nothing on those who have installed, maintained, calibrated and repaired their instruments or performed observations and recorded information in the heart of the observatories. How could we obtain data, photographs, measurements without them?

The National Astronomical Observatory of Chile was fundamentally organized around its hierarchies: first astronomer, second astronomer, collaborators and technical staff.[3] Where one was situated on this ladder determined not only one's remuneration but also one's prestige and, above all, one's visibility. Is this what makes it so hard to uncover the traces of these secondary figures, who escape from us amid the fame of the astronomers, their names and discoveries? Here we must follow paths that we often believe to be far removed from those of science. If, as some have argued, it was the humble people, those who go unrecognized in the history books,

---

[1] Keenan et al. (1985).

[2] Wüst (1913). The newspaper *La Razón* was a publication associated with the Radical Party of Chile, which espoused a progressive and social-democratic ideology and reflected the discourse surrounding the workers' and students' movements. See Santa Cruz (2003).

[3] See Keenan et al. (1985).

who anonymously laid the foundations for knowledge, then we can find traces of their role in rarely-seen documents, those that go almost unperceived and are far from interesting to researchers who only seek out the "Great Men of Science."[4] It is here where work permits, administrative briefs, expense reports, decrees, legal documents and other records cataloged as "unscientific" become important to finally uncovering another history. Going still further, we should pay attention to those destabilizing moments in which tasks fail to be performed or are performed incorrectly in the eyes of the astronomers. It is precisely these "corrective" actions that strip astronomical work down to its essence: everyday labor, trial and error, adaptations. Only in this direction can we find "the voices of the voiceless," carefully selecting records in search of always-scarce evidence. This does not involve, as some have argued, "salvag(ing) some peripheral aspects of the development of science," but rather justly demonstrates how "the scarce evidence illuminates the hidden core of that development."[5]

The argument made by the precision mechanic Richard Wüst in the early twentieth century reveals that ignorance of technical and collaborative work was simply a reflection of their valorization at the time. This explains the difficulty in finding records of the work of technicians and collaborators. Workers go nearly unmentioned in the publications of astronomers, including their memoirs, in part because they themselves didn't see their contribution to the creation of scientific knowledge. Those who published deserved all the credit, in other words. Furthermore, as some recent research has mentioned, while historical and sociological studies of science profusely refer to the notion of the production of knowledge, they use the term "production" in a metaphorical sense: as a collective, organized undertaking "as contrasted with the view of science as an activity conducted by 'great scientists.'"[6] Investigators who study the "production of scientific knowledge" also tend to be more interested in the flexibility of local practices than in the patterns of repetitive actions found in routine work.[7] How can we cut through this epistemic knot that condemns us to repeat the same narrative over and over? What happens if we situate ourselves beyond the idea of scientific progress and its achievements that is so common in the history of science? What will we find by studying the unglamorous stumbles and errors of routine work? As certain studies have emphasized, the role of everyday scientific work remains nearly invisible "unless something goes wrong: instruments do not perform as expected, reagents are impure, standards are contested."[8] As we shall see in this book, controversies, conflicts and failures allow us to pull back the thick veil that has hitherto hidden all but the leading figures.

---

[4] Conner (2005, 4).

[5] Conner (2005, 4).

[6] Gaudilliere and Lowy (1998, 4).

[7] Gaudilliere and Lowy (1998, 345).

[8] Gaudilliere and Lowy (1998, 4).

1.1   The Social Origins of Science

We know that science is a social activity, yet when we study it, it's as if there was a frontier between knowledge and its conditions of possibility, what some have called the *internal* and *external* aspects of science.[9] It's as if those who study the internal aspects of science, who are often scientists themselves, problematize those issues that are *strictly scientific*, while those who work on external aspects, such as historians or sociologists, problematize the *social circumstances* that surround scientific issues. To give one example, astronomers discuss the advancement of ideas about the universe and historians discuss the difficulties faced by these ideas when they come into conflict with institutions such as the church. It's true that this vision has been questioned in the history and sociology of science, but these assumptions have solidified a tradition that has given scientists a leading role in the creation of knowledge. The consequences of this vision include the idea that science may be affected by its surrounding social and cultural circumstances, but that, deep down, rational scientific discourse is an atemporal reality and can be reduced to its rational aspects accumulated over time in a vision of continuous progress. The labor conditions of scientific work, as well as the remuneration for technical personnel and other forms of compensation, would therefore seem to have little relevance. That scientists are often studied individually rather than as part of working groups has done much to further this image. Nevertheless, we have all seen how the crisis provoked by the COVID-19 pandemic revealed that it would have been impossible to successfully treat an unimaginable number of patients without technical personnel (surgical assistants, medical technicians, nurses, assistants, etc.), as it's not enough to simply have doctors and the latest-generation technologies. As Wüst argued over one hundred years ago: all the *day laborers of science* are equally important.

How social can science be? Can we go beyond the leading role of academics and scientists? Is it possible to overcome the vision of the internal and the external? Edgar Zilsel has been key to reflections on what he called "the social origins of science," which later came to be known as the "Zilsel thesis."[10] This "thesis" sought to offer a causal, historical/sociological explanation of the origins of modern science in the Western world between 1300 and 1600. According to Zilsel, modern science emerged as a consequence of the economic, political and technological needs of the new capitalist society, which "required the interaction between three cultural groups that had been socially separated until then: the scholastic university scholars, the humanists, and the artisans." This connection allowed for "the integration of the different theoretical and practical skills mastered by each group: the methodical and logical intellectual training of humanists and scholars, on the one side, and the experimental, causal thinking and the quantitative method of the craftsmen on the other side."[11] What was the impact of these ideas?

---

[9] Latour (1983).

[10] The idea of the Zilsel thesis was institutionalized by Steven Shapin (1981, 450). See: Romizi et al. (2022, 14–17).

[11] Romizi et al. (2022, 15).

From the start, Zilsel's ideas were interpreted by scholars as an external perspective on science. From the 1950s on, Zilsel's conception of the history of modern science had its counterpoint in Alexandre Koyré's interpretation of the scientific revolution, understood "as a genuine epistemological interpretation,"[12] while Zilsel's thesis was characterized "as a purely sociological thesis that does not contribute much to the epistemological understanding of the production of scientific knowledge."[13] This simplification of both positions has led to an underestimation and even erasure of the epistemological aspects of Zilsel's work. In recent years, however, a revisionist perspective has emerged on the Austrian thinker, arguing that Zilsel offers not only a sociological but also an epistemological approach. In other words, that he offers a social vision of science with an underlying epistemology.[14]

Does this epistemology allow us to better understand the role of the *day laborers of science*? Zilsel argued that the new artifacts of the sixteenth century gave rise to modernity, but not as the product of a theoretical and metaphysical attitude that transformed the imaginary, as Koyré emphasized, but to a large extent as a "technological"—that is, "artificial"—product.[15] If we take up Zilsel's ideas and assume that knowledge depends on the dominant societal model, then the study of labor relationships within scientific institutions takes on another meaning.[16] Zilsel's thesis allows us to understand the National Astronomical Observatory not as the product of the individual drives of its scientists, but rather as a joint activity involving those who calibrated, maintained and installed its instruments and telescope huts, those who performed nighttime observations and daytime calculations and the astronomers themselves. Are the latter, located at the summit of knowledge produced at observatories, immune to issues of telescope maintenance or ways of recording observations and collecting and processing data? Was there support among technicians and collaborators when adapting astronomical concerns to local observation conditions? If this is the case, how can we study these connections? Zilsel's epistemological principle, as has been emphasized, involves understanding that knowledge is the collective product of the many interactions, activities and occupations of different practitioners in their social and historical context.[17] The precise word used by Zilsel is cooperation: "This means that science, both in the theoretical and the utilitarian interpretation, is regarded as the product of a co-operation for non-personal ends, a co-operation in which all scientists of the past, the present and the future have a part."[18]

---

[12] Condé (2022, 267). To understand Zilsel's role in the labor history perspective of science, see Roberts et al. (2023) and Hui et al. (2023).

[13] Condé (2022, 269).

[14] Condé (2022, 269).

[15] Condé (2022, 273).

[16] Condé (2022, 273).

[17] Condé (2022, 283).

[18] Zilsel (2003, 220–225).

Studying knowledge and science as an essentially collective phenomenon rather than as the temporal succession of scientific discoveries allows us to sideline questions of the *internal* and the *external*.[19] Zilsel, rather than establishing a sociological thesis, sought to understand the epistemological mechanisms for the social construction of scientific knowledge.[20] Just as the Austrian philosopher understood the rise of modern science as a phenomenon involving the intervention of other historical agents (artisans, sailors, blacksmiths, potters, clockmakers and builders of musical instruments), this book seeks to understand how manual labor and practical knowledge contributed to scientific knowledge.[21] That we often cannot see the practical knowledge of technicians and collaborators does not mean that they did not contribute to the scientific production of astronomers. Here, Zilsel serves as a guide rather than a fixed system. It has been argued that the vagueness of the "Zilsel thesis" has allowed it to be used for interpreting the social basis of science from different perspectives.[22] This book makes it its own.

1.2   Technical Work and Materiality

The work of technicians, builders and collaborators supposes that material mediation is the condition through which, under specific circumstances, the instruments and other objects within an astronomical observatory can function. But to make this relationship clear, it's necessary to abandon the notion that material culture occupies a marginal place in society. Agreements, protocols and norms are materially mediated. This means that, no matter how much we attempt to delineate those workspaces that allow for precision and calibration, we are dependent on what Friedrich Kittler has called the material presence of an object and its practical function.[23] Following Kittler, in order to make this epistemic situation visible, it's necessary to suspend the idealist premise that knowledge, psyche or spirit manage or control that which is beyond us—that is, objects.[24] Kittler's argument, with its roots in Martin Heidegger, provides the condition of possibility for resituating the study of materiality and, at the same time, leaving behind the narrative of the subject as the hero of knowledge and master of modernity.[25] Our philosophical and historiographic tradition is based on the hypothesis that spiritual and immaterial forms of expression—language and the text—are the most elevated,[26] while a society's material forms of expression, such as craft and technique, are perceived as minor issues. It is then assumed that

---

[19] Zilsel (1942).

[20] Romizi et al. (2022, 269).

[21] See: Condé (2022, 272).

[22] Romizi et al. (2022, 269).

[23] Kittler (2000).

[24] See: Kittler (2012, 416).

[25] See: Heidegger (2000, 168).

[26] This has been called the "dictatorship of human being." Olsen (2013, 11).

material objects are ultimately external or banal: it's only worth investigating the spiritual principles behind the objects.[27] The hegemony of these positions has made it difficult to valorize the material world and the role of technicians, as well as the fact that our existence is based on things.[28] We should pay much closer attention to the material components that constitute the condition of possibility for everything we associate with power and social order.[29]

Turning to material mediation, we come across the problem of technique. Under David E. Wellbery's interpretation of Kittler, media and their materiality are determined by the technological possibilities of the time. This position, which some have called the "presupposition of mediality," encourages us to rethink the very notion of technology.[30] A question arises that David Edgerton has taken to the very limit of our preconceptions: Should we continue utilizing notions associated with novelty, the future, invention and creation? Can we carry on with our chronologies of the history of technology based on dates of innovation and its diffusion toward the periphery?[31] A history of the work at an astronomical observatory like the one in Chile would thus consist of a list of its imported technologies and devices: it would be enough to indicate the role they played and their places of origin in order to have an idea of how advanced its astronomy was.[32] Yet if we wish to observe what we haven't seen until now, we must shift our perspective.

What do we find when we undertake a history of the *technology in use*? For Edgerton, this opens up the possibility of an entirely different conception: it reveals to our eyes "a whole invisible world of technologies" and "even more importantly it alters our picture of which have been the most important technologies."[33] This is precisely the central question of this book, which seeks to understand technical and collaborative work in the early history of astronomy in Chile instead of studying innovation processes or the use of new technologies. Stabilizing technological objects is precisely the work of the *day laborers of science*. Stabilization requires instruments to be installed and calibrated: in other words, that they be *used properly*. Disciplinary practices, instruction manuals and norms were not enough. If a telescope didn't look where it was supposed to, if its lenses were defective or improperly installed, if the hut didn't have the required mobility, then none of the data produced would enter the circuits of global knowledge. There would be no astronomy done, no matter how much was invested in the biggest, most innovative and advanced telescopes.

---

[27] Hahn (2014, 9-11).

[28] Olsen (2013, 4).

[29] Olsen (2013).

[30] "The decisive methodological step undertaken by Kittler is to generalize the concept of medium, to apply it to all domains of cultural exchange. Whatever the historical field we are dealing with, in Kittler's view, we are dealing with media as determined by the technological possibilities of the epoch of question. Mediality is the general condition within which, under specific circumstances, something like 'poetry' or 'literature' can take shape." See: Wellbery (1990, XIII).

[31] Edgerton (1990).

[32] The most popular history of the National Astronomical Observatory of Chile engages in this very exercise. See Keenan et al. (1985).

[33] Edgerton (1990, 18).

There's a major obstacle to studying the technologies in use, or what Edgerton himself calls "creole technologies" associated with local knowledge: they often go unseen, leaving behind no record. This situation raises a major epistemic and methodological challenge: how can we study the *day laborers of science* when technicians and collaborators didn't leave behind letters and diaries, when there's almost no trace of their activities in the press? At most, as the National Astronomical Observatory of Chile is a government institution, they appear in contracts and records of payments to construction workers, solderers and clockmakers: but how can we trace their role, their influence? How can we study what Klaus Hentschel has called *the invisible hand*?[34]

The thesis that the work of technicians has not been sufficiently emphasized has a long history in specialized studies. Ever since Steven Shapin's seminal article *The Invisible Technician*, historians of science have become more conscious of the importance of technicians, "who cooperated closely with scholars in making and recording all kinds of scientific knowledge."[35] Shapin speaks of a double invisibility of technicians and collaborators: both to those who left behind records of scientific activities at the time (who made no more than a handful of references) as well as to the scholars who later sought the traces of history in these records.[36] Shapin himself notes that, when scrutinizing the archives in detail, there's times when we can nevertheless discern that technicians occasionally assumed full responsibility for conducting experiments, yet this was not reflected in the texts written by those leading the investigation or, in our case, in the observations and measurements published by astronomers. These difficulties with the records will be explored over the course of this book.

Broadening the spectrum of scientific production isn't simply a question of justice: expanding our notion of knowledge allows us to give voice to those carrying out essential tasks, which have gone largely unheard until now, but it also allows us to paint a more complex picture of *doing science*.

As Shapin himself has argued, it allows us to stop visualizing knowledge and scientific work as something individual. If we pay close attention to what the instruments tell us, as well as how they're handled, then maybe we'll come to understand that what we call research groups are not mere sums of individuals. For astronomical observations to be conducted under the parameters accepted by scientific communities, they need to function as a team. If we examine science beyond the predominant tendency of *thought* and the individual, and instead approach it through the notion of *labor*, we will paint quite a different picture. Hentschel argues that this interest in the individual expresses more than mere vanity or the public's fascination with geniuses and "great" personalities, as it also implies an image of science and of knowledge itself: "Most of the time, this genius is also a 'theoretician,' while experimenters and technicians are only given the role of an executive organ under this

---

[34] Hentschel (2008).

[35] Zuidervaart (2012, 59).

[36] Shapin (1989, 554–563).

'distorted image.'"[37] As in the Zilsel thesis, studying technical work not only leads us to question the notion of the subject as the protagonist of knowledge, but also the idea that all scientific production is the exclusive product of rationality. When we stop to observe the activities of technicians, we see the role played by improvisation in the construction of knowledge to the extent that it is the "main fuel" of maintenance and repair workers, whose interventions always "overwhelm the standardized procedures."[38] Chance, a crucial factor and yet one absent from history, takes on a new meaning here.

Studying the relationship between technicians and materiality opens up a new path, albeit one not free from pitfalls: that of studying actions for the maintenance and repair of the technologies in use.[39] It is precisely during moments of imperfection and maladjustment when the technicians and mechanics in charge of maintenance and repair become visible and play a leading role. This cloak of invisibility can be explained by both the social situation of the technicians themselves as well as by conceptions of scientific work and the production of knowledge.[40] By separating intellectual labor from manual labor, Western history has underestimated the latter, denying it any possibility of being recognized as a valid generator of knowledge.[41] It's therefore not a surprise that technicians only rarely appear in scientific discourse and the history of science. There are nevertheless certain situations in which technicians become visible, such as laboratory accidents, failed experiments or disputes between technicians and scientists. Both disputes as well as maintenance and repair work can serve as a door to understanding and evaluating the work of technicians, as will be seen in this book.[42]

Though instruments often go unperceived in everyday scientific routines, imperfections and breakage disarticulate scientific work and repairs need to be made for science to resume its normal course.[43] Maintenance and repair therefore serve as a hinge between the inevitable phenomenon of breakage and imperfections and the smooth functioning of scientific practice.[44]

Despite its invisibility, the problem of maintenance is crucial as the very nature of technology rests on a basis so fragile that any fault could provoke the breakdown of the entire system. To the extent that technologies require greater supervision and ever more meticulous maintenance, as with the precision instruments examined herein, their maintenance and repair occupies a large part of scientists' time. These realities

---

[37] Hentschel (2008, 19).

[38] Denis and Pontille (2015, 355).

[39] Edgerton (1990), Schaffer (2011, 706–717). See also: Denis and Pontille (2015, 2017) and Russell and Vinsel (2018).

[40] This was the argument of Shapin (1989).

[41] Morus (2016).

[42] Shapin (1989) and Morus (2016).

[43] Graham and Thrift (2007) and Jackson (2014). This problem has been profusely addressed in studies on technological trajectories. See: Dosi (1982, 147–162), Cohen and Wesley (1990) and Viotti (2002).

[44] Jackson (2014).

rarely take up many pages in the history of science: "mundane and infuriating, full of uncertainties, they are among the major annoyances surrounding things."[45]

The social order rests on the material fragility of things; thus the constant need to take care of them. Maintenance and repair are therefore a necessary condition for all stabilization efforts, given that they reveal that this order (which is also *socio-material*) is the concrete product of everyday practices for the maintenance and repair of the material. Emphasizing the instability, fragility and possible failures that constantly threaten this order offers us an opportunity to reconsider the traditional role of technicians in society and, more generally, the agency of objects.[46]

1.3   Stabilizing the Local

Astronomical observatories look the same from a distance, as if they were all built at the margins of the site.[47] If one observes their domes, their architectonic layout and the forms taken on by their telescopes, there's a deliberate attempt to build similar structures. Where does this interest come from? Is this a question of influences and styles or does it correspond to a search for a uniform model? And if that's the case, what is being emulated? The copy, the replicability of observations and control over procedures for recording data all form an essential part of the circuit of modern scientific knowledge, which naturally includes astronomy.[48] We often ignore the efforts made by disciplines to achieve international validation. The very notion of objectivity has been built on uniformity: it allows for the articulation of global networks, which are necessary for the construction of knowledge, as everyone is doing the same thing, making use of similar techniques and instruments.

In this attempt at unification, astronomers employ strategies to erase or silence local specificities in what Peter Galison has called the *local delocalization* of scientific work.[49] And yet local reality appears when it's least desired: when instruments break down, when there's nobody who knows how to work them properly or in the precise moment in which an observation cannot be made because of a poorly calibrated chair or an immobile dome. No matter how hard astronomers work at building a *nonplace*, local destabilization threatens to ruin nights of work and enormous investments.[50]

It has been said that astronomy was one of the first precision sciences, but how was it constructed at the intersection of the local and the delocalized? How can we even begin to unravel this knot?

---

[45] Edgerton (1990, 102).

[46] Denis and Pontille (2015, 8).

[47] Part of these reflections have appeared in Sanhueza-Cerda (2022).

[48] On replicability, see: Collins (1985). On the copy, see: Edgerton (1990). On observation, Daston (2008).

[49] Galison (2005, 409). Taken from Nasim (2017, 181).

[50] Nasim (2017, 181).

One answer is to forget everything we think *happens* at an observatory. Rather than understanding that it's where one observes the skies and measures the stars, we can think of it, above all, as "a workshop where a wide range of technological devices—optical instruments, electrical apparatuses for telegraphy, clocks—were developed, tested, calibrated, and put to extensive use."[51] This enterprise, which we can understand as a factory of knowledge, brings together astronomers with instrument manufacturers, but also architects, technicians, construction workers and maintenance personnel, not to mention those who transcribe data from the instrument to a visual format, tables and publications, as well as those who make the necessary calculations on devices specially designed for the transfer of information.

A second approach involves looking *beyond* the domes of observatories. Observation techniques, defined as a set of necessary practices for successfully executing the ocular vision of the telescope, are frequently understood as a question of methods, protocols and routines employed by astronomers.[52] Yet not everything occurs within the observatory. What occurs outside its walls forms an integral part of these practices, including the administration of the budget, the construction and installation of instruments and buildings, the hiring of staff, maintenance and repair, as well as the prestige and valorization of the institution by public opinion. For the astronomical observatory to ensure "the calibration, manipulation, and coordination of precision instruments for making observations and taking measurements,"[53] it's necessary for its walls to disappear.

What do we see when we approach what an observatory *is*, as well as *its* limits, without preconceptions? We know that observatories seek the controlled manipulation of observation and measurement instruments. Is that not what definitively occurs behind and within its walls? If this is the case, we would then have to ask what makes everything function. In other words: How is precision constructed? Above all, we must examine how precision is understood through the stabilization of objects as a way of making visible the everyday labor that sustains it.

David Aubin has argued that, in recent decades, historians of science have studied the ways in which scientific knowledge is locally constructed in a specific place and under specific circumstances. Insertion in a local environment can have effects on the types of knowledge produced and, therefore, on the nature of the activities conducted there.[54] Aubin and Stéphane Le Gars have developed the idea that astronomical observatories construct a place of *de-placement*. The strategies used at observatories are therefore far from those used in the laboratory sciences: while laboratories attempt to become *nonplaces* and distance themselves from local conditions as much as they can, observatories cannot employ this strategy because they need these conditions to be taken into account in the outside world.[55] The transformation of the observatory into a standard data point within observation networks implies a relative lack of

---

[51] Aubin et al. (2010, 10).

[52] Aubin et al. (2010, 6).

[53] Aubin et al. (2010, 6).

[54] Aubin (2017).

[55] Le Gars and Aubin (2009, 509–531).

place, or a delocalized space in Galison's words, but at the same time demands that the local be incorporated into the global. In this way, the data from an observatory can be correlated with that from elsewhere. In the *laboratory-field* dyad, the place for the production of astronomical knowledge is a heterotypical space: a place in a network, yet *de-placed*.[56]

So that periodic observations can function within this network, it's necessary to neutralize observations. In other words: to make instruments function correctly at this intersection between the situated and the *de-placed*, the local and the global. Omar Nasim argues that an observatory "is as stable as its primary instruments" and so everything must be done to stabilize them, whether from a cultural and epistemic perspective or, above all, materially.[57] For the neutralization of stabilization to work, it must be framed by what Aubin calls a *regime of spatiality*. This means that the instruments are situated on three levels: in the architecture of the building, in relation to their layout within the observatory and in the nature of the geographic and abstract space in which they are found.[58] This explains why these places of science must constantly adapt to the size and layout of their objects of observation and measurement. For example, changes in transit instruments during the seventeenth century, in which they became smaller and more unstable, forever modified the architecture of the observatory. New telescopes began to be set in brick or granite pillars that were sunk into the foundations of their observatories and sought to avoid vibrations through the use of piers. Telescopes could no longer be placed on university rooftops or church spires, as was habitual until then, and so new sites were sought out that were far from urban environments and located at a considerable height. At the same time, it became necessary to construct movable domes that would provide the telescopes with shelter and visibility. As telescopes achieved greater levels of precision, the stability of instruments became a condition for astronomical observation practices.[59]

It is said that observation dynamics serve the precision and exactitude of their instruments.[60] Precision thus took on the characteristics of an "action-at-a-distance technology" by stabilizing the truthfulness of the information registered at observatories.[61] This technology cannot be reduced to the instrument itself: it is a set of protocols that function on the basis of networks of observers who must learn "to adjust their experiences, their ways of writing and acting (...) so that the movements and displacements that take place at the observatory (...) are transparent."[62] Having a precision instrument did not ensure its use within the transnational network. If

---

[56] Le Gars and Aubin (2009, 513–520).

[57] Nasim (2017, 181).

[58] Aubin (2017).

[59] Nasim (2017, 181).

[60] From the start, astronomy has been defined as "the oldest of the exact sciences," invoking the invention of the telescope in 1608. It has even been argued that this discipline represents the *materialization* of science, overcoming the old Greek division between *theoria* and *techne* through the precision of the artisan wedded to an abstract vision of the cosmos. See Müller and Posch (2010, IX).

[61] Pérez (2010, 33).

[62] Pérez (2010, 34).

its data could not be protocolized, "it was relegated to fulfilling a purely rhetorical function." The physical handling of the instrument must be subjected to a "communitarian objectivity" that would guarantee "homogeneity of results."[63] The calibration and control of instruments is therefore crucial, as it guarantees the circulation of the knowledge they produce. At the same time, problems in establishing precision become questions of community accords and, therefore, standardization.[64] For this to occur, there needs to be agreement on the standards for comparison. Precision, as M. Norton Wise has said, "is always the accomplishment of an extended network of people."[65] This requires a consensus on materials, instruments, methods and values. Modes of precision are thus a joint product of social labor and it is precisely the labor dimension of science that is the theme of this book.

References

Aubin, D. 2017. L'observatoire: Régimes de spatialité et délocalisation du savoir, 1769–1917. In *Histoire des sciences et des savoirs de la Renaissance à nos jours*, ed. Dominique Pestre, vol. 2, 54–71. Paris: Le Seuil.

Aubin, D., C. Bigg, and H. Sibum, eds. 2010. *The heavens on earth: Observatories and astronomy in nineteenth-century science and culture*. Durham, London: Duke University Press.

Cohen, W. M., and D. A. Levinthal. 1990. Absorptive capacity: A new perspective on learning and innovation. *Administrative Science Quarterly* 35 (1): 128–152.

Collins, H. 1985. *Changing order replication and induction in scientific practice*. Chicago: University of Chicago Press.

Condé, M. L. 2022. The epistemological foundations of the Zilsel thesis. In *Edgar Zilsel: Philosopher, historian, sociologist*, ed. D. Romizi, M. Wulz, and E. Nemeth, 267–284. Cham: Springer.

Conner, C. D. 2005. *A people's history of science: Miners, midwives, and low Mechanicks*. New York: Bold Type Books.

Daston, L. 2008. On scientific observation. *Isis* 9: 97–110.

Denis, J., and D. Pontille. 2015. Material ordering and the care of things. *Science, Technology, & Human Values* 40 (3): 338–367.

Denis, J., and D. Pontille. 2017. Beyond breakdown: Exploring regimes of maintenance. *Continent* 6 (1): 13–17.

Dosi, G. 1982. Technological paradigms and technological trajectories: A suggested interpretation of the determinants and directions of technical change. *Research Policy* 11 (3): 147–162.

Edgerton, D. 1990. *The shock of the old: Technology and global history since 1900*. London: Profile Books Ltd.

Galison, P. 2005. Material culture, theoretical culture, and delocalization. In *Theatrum Scientiarum: Collection, Laboratory*, ed. H. Schramm, L. Schwarte, and J. Lazardzig, 490–506. Theater: Berlin.

Gaudilliere, J. P., and I. Lowy. 1998. *The invisible industrialist: Manufactures and the production of scientific knowledge*. New York: St. Martins Press.

Graham, S., and N. Thrift. 2007. Out of order understanding repair and maintenance. *Theory, Culture & Society* 24 (3): 1–25.

---

[63] Pérez (2010, 39).

[64] Wise (1995, 8).

[65] See: Wise (1995, 9).

Hahn, H. P. 2014. *Materielle Kultur. Eine Einführung*, Reimer: Berlin.

Heidegger, M. 2000. Das Ding. In *Vorträge und Aufsätze, Part II*, 3, vol. VI. Frankfurt/M: Vittorio Klostermann.

Hentschel, K. 2008. *Unsichtbare Hände, zur Rolle von Laborassistenten, Mechanikern, Zeichnern u. a. Amanuenses in der physikalischen Forschungs- und Entwicklungsarbei*, Diepholz; Stuttgart; Berlin: Verl. für Geschichte der Naturwissenschaften und der Technik.

Hui, A., L. Roberts, and S. Rockman. 2023. Introduction: Launching a labor history of science. Focus: Let's get to work: Bringing labor history and the history of science together. *Isis* 114 (4), 817–826.

Jackson. 2014. Rethinking repair. In *Media technologies: Essays on communication, materiality, and society*, ed. T. Gillespie, P. J. Boczkowski, and K. A. Foot, 221–239. Cambridge/Massachusetts: MIT.

Keenan, P. C., S. Pinto, and H. Alvarez. 1985. *El Observatorio Astronómico Nacional de Chile (1852–1965)*, 1985. Santiago: Universidad de Chile.

Kittler, F. 2000. *Eine Kulturgeschichte der Kulturwissenschaft*. Munich: Fink.

Kittler, F. 2012. *Die Wahrheit der Technische Welt*. Berlin: Suhrkamp.

Latour, L. 1983. Give me a laboratory and i will raise the world. In *Science observed: perspectives on the social study of science*, ed. C. K. Knorr and M. Mulkay, 141–170. London: Sage.

Le Gars, S., and D. Aubin. 2009. The elusive placelessness of the Mont-Blanc observatory (1893–1909): The social underpinnings of high-altitude observation. *Science in Context* 2009 (22): 509–531.

Morus, I. 2016. Invisible technicians, Instruments-makers and Artisans. In *A companion to the history of science*, ed. B. Lightman, Bernard, 95–110. New York: Willey-Blackwell.

Müller, H. J., and T. Posch. 2010. *Die Geschichte der Sternwarte Wien*. Frankfurt am Main: Verlag Harry Deutsch.

Nasim, O. 2017. Observatorium. In *Handbuch für Wissenschatsgeschichte*, ed. Marianne Sommer, 180–192. Stuttgart: J. B. Metzler Verlag.

Olsen, B. 2013. *In defense of things. Archaeology and the ontology of objects*. Lanham: Altamira Press.

Pérez, N. 2010. *Actos de precisión. Instrumentos científicos, opinión pública y economía moral en la Ilustración española*. Madrid: CSIC, 2007.

Roberts, L., S. Rockman, and A. Hui. 2023. Historiographies of science and labor: From past perspectives to future possibilities. *History of Science* 61 (4): 448–474.

Romizi, D., M. Wulz, and E. Nemeth. 2022. *Edgar Zilsel: Philosopher, historian, sociologist*. Cham: Springer.

Russell, A. L, and L. Vinsel. 2018. After innovation, turn to maintenance. *Technology and Culture* 59 (1), 1–25.

Sanhueza-Cerda, C. 2022. Stabilizing local knowledge: The installation of a Meridian Circle at the National astronomical observatory of Chile (1908–1913). *Isis* 113 (4): 710–727.

Santa Cruz, E. 2003. El campo periodístico en Chile a principios del siglo XX. *Periodismo y Sociedad* 14: 17–29.

Schaffer, S. 2011. Easily cracked. Scientific instruments in states of disrepair. *Isis* 102 (4), 706–717.

Shapin, S. 1981. In *Dictionary of the history of science*, ed. W. Bynum et al., 450. London/Basingstoke: Macmillan.

Shapin, S. 1989. The invisible technician. *American Scientist* 77: 554–563.

Viotti, E. 2002. National learning systems: A new approach on technical change in late industrializing economies and evidences from the cases of Brazil and South Korea. *Technological Forecasting and Social Change* 69 (7): 653–680.

Wellbery, D. E. 1990. Forward in F. Kittler, *Discourse networks 1800/1900*, XIII. Sandford: Stanford University Press.

Wise, M. Norton. 1995. *The values of Precision*. Princeton, N.J.: Princeton University Press.

Wüst, R. 1913. Wüst, R., "Lo del Observatorio Astronómico" (letter from Ricardo Wüst), News Paper *La Razón*, Santiago de Chile, 3 Jan 1913.

Zilsel. 1942. The sociological roots of science. *The American Journal of Sociology* 47, 544–562.

Zilsel, E. 2003. *The social origins of modern science*. Dordrecht: Springer.

Zuidervaart, H. J. 2012. The 'Invisible technician' made visible: Telescope making in the seventeenth and early eighteenth-century Dutch Republic. In *From earth-bound to satellite: Telescopes, skills, and networks*, ed. A. D. Morrison-Low, et al., 59. Leiden/Boston: Brill.

Chapter 2
Astronomy, Labor and Working Conditions in Chile

Abstract The chapter addresses the relationship between science, labor, and working conditions. It points out that knowledge production in astronomy has historically depended on a hierarchical labor structure separating principal investigators, technicians, and administrative staff, relegating technical and manual work, such as mechanics and assistants, to invisible roles. In the Chilean context, developing a working class in the nitrate and railway industry in the nineteenth century promoted the expansion of technical knowledge. The lack of skilled labor led to 'on-the-job' training and collaboration with foreign technicians, albeit with conflicts. Integrating a working class into science reflected an effort to achieve a culture of progress and technical knowledge. This established a basis for professionalization in astronomy and other sciences in Chile, with the state playing a crucial role in the training of technicians and the modernization of the industry.

Keywords Labor history of science · History of labor in Chile · Mid-nineteenth century · Early twentieth century

For some years, there has been a renewed interest in studying science from the perspective of labor history. Historians such as Simon Schaffer and Steven Shapin have pointed out how laboratories and observatories were also places of physical and technical work, reconfiguring the idea of the scientist as a laborer.[1] This approach has expanded in the field of the history of astronomy, especially in the work of historians such as Deborah Jean Warner, who has emphasized the importance of manual and technical work in the history of scientific instruments and observational methods in the American context.[2] In this sense, the analysis of astronomy from a labor perspective not only recovers the contributions of those who worked on the margins of the field (such as craftsmen, technicians, and assistants) but also allows for a more comprehensive understanding of how scientific knowledge is the result of a collective process. This opens the door to new discussions about the valuation

---

[1] Schaffer and Shapin (1985).

[2] Warner (1968).

of non-intellectual labor in science and how professionalization and specialization have impacted on the labor structure within science.

The study of the labor structure of science—contracts, work expectations, division of labor, and salaries—is fundamental to understanding how scientific knowledge is produced in its social and institutional context. Examining science from a labor perspective highlights that knowledge production is not a neutral or merely intellectual process but is influenced by scientists' labor and economic conditions. The socio-political contexts that shape scientific work can reveal the power structures inherent in knowledge production. Hence, issues such as exploitation, precarity, and the impact of capitalism and colonialism on scientific practices are crucial to understand. It has even been suggested that labor history should be merged with the history of science because such integration can provide a complete understanding of scientific practices.[3] At the same time, hierarchical structures within scientific institutions delineate the responsibilities of different groups: principal investigators, assistants, technicians, and administrative staff. This division of labor organizes scientific production and reflects structural inequalities in the distribution of prestige, recognition, and financial rewards. For example, in the history of astronomy (as will be discussed later in this book), the relationship between the mechanics (who made the instruments work) or the 'human computers' (who performed labor-intensive calculations) and the astronomers who managed the projects, demonstrates how manual and technical labor has been central to scientific progress, but has often been invisible in the mainstream narrative. It has been pointed out how gender and class hierarchies have played a significant role in the division of scientific labor, with women and other marginalized groups often relegated to positions of lower prestige.[4] At the same time, this division of labor exposes the *caesura* between core and peripheral countries, which, in turn, also reflects and perpetuates power dynamics and hierarchies within the scientific community itself.[5]

For centuries, astronomical observation depended on using rudimentary instruments and the meticulous work of astronomers, who were often also craftsmen of the very instruments they used. In the Renaissance, figures such as Tycho Brahe combined observational practice with technical work and the construction of observatories, showing how the production of scientific knowledge was intrinsically linked to manual and technical work. Brahe's Uraniborg observatory, for example, was both an instrument-making workshop and a center for astronomical observation, highlighting the importance of labor in generating astronomical data.[6] Furthermore, the role played by telescope builders, such as the Repsold brothers in Hamburg in the nineteenth century, reveals the extent to which working with scientific instruments became a criterion for validating new knowledge. Accepting the validity of the telescope often meant accepting the validity of the discoveries.[7] At the same time, the

---

[3] Hui et al. (2023).

[4] Schiebinger (1999). For the case of Chile, see Sect. 6.3 in this book.

[5] Belteki (2023).

[6] Thoren and Christianson (1990) and Christianson (1999).

[7] King (1955), Dunn (2009) and Chinnici (2017).

importance of manual labor and craftsmanship in astronomy has been emphasized, especially in drawing, as an essential tool for observation and knowledge production.[8]

In the nineteenth century, with the creation of national observatories in Europe and the Americas, astronomy became a more clearly defined profession, and scientific work became even more specialized. The role of 'human computers,' such as the women working at the Harvard Observatory under Edward Charles Pickering, is a clear example of how the manual labor of analyzing astronomical data remained vital to scientific progress. These women, as well as those we will see in this book, performed essential calculations to classify stars and perform spectrographic analysis, which stresses that the production of astronomical knowledge was, in many ways, a collective process of technical work.[9]

2.1 Chile's Labor History

To fully grasp the significance of *The Day Laborers of Science*, it is essential to explore the history of technical work in Chile, particularly during the late nineteenth and early twentieth centuries, within the framework of national expansion and the consolidation of industrialization. Historical analyses of labor contracts, the division of labor, and precarious employment conditions unveil the power dynamics that shaped the production of scientific knowledge in Chile. These structural factors are mirrored in labor history outside of scientific institutions, emphasizing the fight for improved working conditions and social rights. This parallels the development of astronomy and other scientific fields. Such perspectives facilitate a more comprehensive understanding of how knowledge production is interwoven with local and global labor and economic contexts.

Beginning in the second half of the nineteenth century, Chile underwent a national expansion process characterized by the incorporation and occupation of new territories through economic and military actions. Both global and local factors drove this process. In the international context, the reduction of transportation costs, the increase in world trade, and the growing demand for raw materials by the industrial powers strengthened the export model of Latin American nations, which tried to extend their borders in search of agricultural and mining resources.[10] At the local level, the decline of silver mining and the end of the wheat cycle in the 1870s redirected economic activity toward the exploitation of minerals, especially nitrates, in the country's north. The strategic importance of this resource was one of the main factors that triggered the so-called War of the Pacific (1879–1883), which resulted in the annexation of extensive regions that had belonged to Peru and Bolivia, rich in nitrate minerals.[11]

---

[8] Nasim (2019).

[9] Mac (1990).

[10] Bértola and Ocampo (2013, 103–104).

[11] Ortega (2005, 405–428).

Within this framework of economic transformations, a new working class formed, representing the transition from the rural to the industrial economy between the late nineteenth and early twentieth centuries. In particular, many historians have investigated the emergence of a working class in the saltpeter extraction of northern Chile, paying attention to the formation of a working-class identity in a context of high labor exploitation. The presence of social conflict has strongly marked this history. It has been argued that as economic structures changed, workers acquired "a new configuration and mode of expression." This, in turn, "led to an open politicization of social relations," a phenomenon that marked a large part of the twentieth century.[12] In this way, a workers' movement emerged in Chile, beginning with the union struggles and the political participation of workers in the productive sector of the nitrates industry and continuing with the formation of a working class in the cities that were becoming more and more industrialized.[13] These changes led to the first massive strikes in the country's north, linked to saltpeter exploitation, which aroused "deep fears" in the conservative sectors, initially associated them with the influence of European socialist and anarchist ideas.[14]

Some key aspects of the Chilean working class are its "idealization of science and technology, the eminently urban and legalistic character of the movement," and the use of the press "as a privileged weapon."[15] Although these were characteristic elements that had been present "throughout the second half of the nineteenth century," they were strongly evident at the beginning of the twentieth century. Indeed, in the first decades of the twentieth century, Chile's insertion into the global capitalist economy meant not only the arrival of imported articles of daily use and capital goods but also the more or less massive arrival of European immigrants (such as the German astronomers and technicians discussed in this book), as well as the entry of English, French and German culture "with the doctrines of social change and progress." The culture that came from Europe was associated with progress for the country and was seen as a model to follow, whether it was real or not. This added to the "appearance of organizations, schools (…) newspapers, and the whole universe of workers' culture that became autonomous".[16] This, in turn, generated a "working culture". This working culture was expressed based on three coordinates: "the forging of an ideology, the consolidation of forms of organization and expression" and the creation of "a working intellectuality."[17] This culture admired science, literature, and art, but at the same time, "it was not the culture of men of science (…) [but rather] manual workers made it". In other words, the working culture wanted to "realize the values of scientific knowledge" by incorporating the place of manual workers in

---

[12] Pinto (1998, 17).

[13] Salazar (2003).

[14] Grez (2007).

[15] Grez (1998, 131–137).

[16] Devés (1991, 127–130).

[17] Devés (1991, 131).

the narrative of progress. This position of working culture expressed an interest in "joining the world of decisions, of power," from which they felt excluded.[18]

2.2 Metal Men: Technical Work in Chile

Technical work in Chile, for the period of this book between the mid-nineteenth century and the beginning of the twentieth century, is closely linked to the nitrate activity in the north of the country and, later, to its impact on the cities of the center, especially in the capital city, Santiago. Chile's industrial expansion was limited from the middle to the end of the nineteenth century, while non-mining metallurgical establishments maintained antiquated structures and practices. This panorama changed drastically with the outbreak of the War of the Pacific against Peru and Bolivia in 1879 and the subsequent beginning of the saltpeter cycle. This period triggered the strategic need to develop a railroad network that guaranteed control of the newly occupied nitrate provinces and, later, connected the center and south of the country. In this way, mining and "a strong railway effort" foreshadowed the installation of a modern industrial metallurgical sector, which became viable "as soon as it became a supplier of equipment and parts for the railroads and mining activity."[19] In this context, non-metallic mining, especially of nitrates and coal, became the engine of industrial growth and strengthened the railway sector, which began to produce equipment and spare parts to meet railway demands.[20]

The so-called "*metal men*" became an almost exclusive technical group in the workshops to repair and manufacture wagons and locomotives. In this context, the old skills of Chilean workers associated with "manual skills" were insufficient, especially in "detailed and precision manufacturing such as machinery and transport equipment," which forced "the use of specialized foreign workers."[21] Regarding technical training, the impact of railroad technology "revealed the country's insufficiency of skilled labor for mechanical work." However, at the same time, it had "a positive influence on the acquisition of new productive capacities by creating standards of technical qualification."[22]

Despite all the changes that the introduction of the railroad in Chile meant, until around the second decade of the twentieth century, the labor force continued to show low qualification indexes. This situation forced the Chilean state to intervene because of the country's productivity crisis. At first, the State sought to regulate the supply and demand of jobs in the agricultural and mining sectors, as well as to provide transportation facilities for nitrate workers in the north of the country, who were experiencing high unemployment after the abrupt fall in nitrate production at the end

---

[18] Devés (1991, 139).

[19] Matus (2009, 14).

[20] Matus (2009, 14).

[21] Guajardo (1993, 161).

[22] Guajardo (1993, 161).

of World War I.[23] However, it was soon realized that it was not enough to move the population from one productive area to another: it was necessary to intervene in their technical training.

How was Chile "able to reach a high density of modern and foreign technology without having, in its beginnings, the labor force and manufacturing centers adequate to operate and sustain such technology?"[24] We know that by the 1930s, Chile reached a railroad coverage of 12 km of track per square kilometer of land area. This situation placed the country above many others in the Americas and Europe. This apprenticeship had begun in the nineteenth century with the training of machinists and mechanics in "skills and technical knowledge destined to operate steam technology." However, it was enormously boosted with the creation of the State Railway Company (EFE) in 1884.[25]

The Chilean railroads, in their beginnings associated with the installation of British capital and industries in the late nineteenth and early twentieth centuries, relied on foreign engineers, mechanics, and technicians, mainly British (although not exclusively). This situation occurred mainly through the operations of British companies such as Anthony Gibbs & Co. However, it could not be sustained by the state-owned Chilean railroad company EFE. It became evident that the country should "take initiatives to stimulate national work" where there were only foreigners.[26]

Initiatives to train technicians in Chile were initially focused on formal education, particularly in the School of Arts and Crafts, which had educated artisans in construction and masonry since the nineteenth century. However, the need for teaching staff in Chile modified the educational scheme. Training on the railroads was part of the production process: technicians, maintenance, and repair personnel would be trained "on the job."[27] This practical training meant training the Chilean *peon* (accustomed to agricultural work) to place him in the worker category. Documentary evidence affirms that this educational process was carried out with the help of foreign technicians and engineers, and was not free of conflicts and accidents. The attitude of disdain and rejection that these foreign experts had towards unskilled Chilean workers was the catalyst that brought about complex process management. In the beginning, this training process consisted of teaching how to clean and polish the oxide of the parts, which meant the first contact with the machines, but above all, a process of disciplining "in the stations, yards and workshops to be able to incorporate into the new productive order and its trades."[28] Later, the first group of Chileans entered the machinist trade at the end of the nineteenth century, encouraged by the Chilean government. This fact was condemned by the English bosses of the machinists' shops, who considered it a lack of respect for their work. It became evident that foreign instructors still found it challenging to consider this advancement of the

[23] Guajardo (1993, 161).

[24] Guajardo (1992, 18).

[25] Guajardo (1992, 18).

[26] Guajardo (1992, 27).

[27] Guajardo (1992, 29).

[28] Guajardo (1992, 31).

Chilean worker to a technical level. Numerous documentary sources show the opposition of foreigners to teaching local workers with the fear of future replacement. Another conflict was the type of contracting and job stability. Many of the *metal men* did not have fixed-term contracts, and the so-called *day laborers* (who worked for a daily salary) made the instruction and technical preparation process very difficult.

This "on the job" model showed its exhaustion in the first decades of the twentieth century, due to conflicts regarding the continuity of the process, the continuous arrival of new technologies, and the high levels of illiteracy of Chilean workers, which prevented them from developing planning or self-training actions based on the study of manuals and protocols. Under the impulse of the state railroad company EFE, the School of Arts and Crafts of Santiago once again took on a leading role in training future mechanics and national technicians. Towards the end of the first decade of the twentieth century, this institution and a group of schools for machinists and workers in the country's main cities were the driving force of the training process. The implementation of new state regulations to control the process of assembling and repairing machines, as well as the implementation of labor practices of the students of the School of Arts and Crafts in the workshops, opened a new panorama for the *metal men*.[29] It was clear to both the state and private companies that training through work experience was the only way to ensure the necessary and sufficient human capital to face technological change. The State of Chile, through its railroad company, was not only a pole of technological development but also stamped a mark on the Chilean technician. In this style, knowledge and practical routines had to be part of their way of confronting work.

2.3 Labor History and Astronomy in Chile

To properly understand the role of mechanics and assistants at the Astronomical Observatory, it is crucial to situate them in the labor context of Chile between the end of the nineteenth century and the beginning of the twentieth century. For example, the context of the "metal men," who operated in the railway and industrial repair and maintenance workshops, was decisive in establishing a technical base in Chile. The growing demand for workers trained in mechanics, instrumentation, and precision techniques prompted the creation of technical training spaces like the School of Arts and Crafts. However, much of the technical learning continued in the workplace, which affected skills development and fostered a work culture based on practical cooperation and direct contact with technology rather than formal academic training.

At the Astronomical Observatory, mechanics and assistants were inserted into a labor hierarchy that reflected these structural divisions. The position of these technical workers was fundamental: they manipulated, repaired, and maintained the observational instruments, making possible the collection of accurate astronomical data.

---

[29] Guajardo (1992, 46).

Despite the importance of their work, their role tended to be invisibilized in astronomical knowledge production accounts, which usually prioritized astronomers and scientists. This marginalization reflects a division of scientific labor similar to that of other technical industries of the time, in which the work of technicians and mechanics, despite its importance, was considered secondary to intellectual work.

In this context, there is a confluence between the ideals of Chilean working-class culture—characterized by admiration for technique and science—and the aspirations of the observatory workers. The training and work in these spaces offered the mechanics a direct connection with scientific progress while allowing them to participate in a culture of technical knowledge that, although not highly visible, was essential for the operation and improvement of astronomy in the country. Indeed, this integration of technicians in scientific processes underlines the idea that knowledge at the observatory was produced collectively, in which manual and technical practice was intimately linked to astronomical research. Therefore, the analysis of the mechanical workers at the Astronomical Observatory not only enriches the understanding of how astronomy developed in Chile but also shows how working conditions, technical training, and the general labor culture of the country somehow (although sometimes not so clearly) influenced the production of scientific knowledge.

References

Belteki, D. 2023. The grand strategy of an observatory': George Airy's vision for the division of astronomical labour among observatories during the nineteenth century. *Notes and Records* 77: 135–151.

Bértola, L., and J. A. Ocampo. 2013. *El desarrollo económico de América Latina desde la independencia*. México: Fondo de Cultura Económica.

Chinnici, I. 2017. *Merz Telescopes. A global Heritage Worth Preserving*. Cham, Springer.

Christianson, J.R. 1999. *On Tycho's Island: Tycho Brahe and his Assistants, 1570–1601*. Cambridge: Cambridge University Press.

Devés, E. 1991. La cultura obrera ilustrada chilena y algunas ideas en torno al sentido de nuestro quehacer historiográfico. *Mapocho* 30: 127–130.

Dunn, R. 2009. *The telescope. A short history*. London: National Maritime Museum.

Grez, S. 1998. 1890–1907: De una huelga general a otra. Continuidades y rupturas del movimiento popular en Chile. In *A noventa años de los sucesos de la escuela Santa María de Iquique*, ed. Pablo Artaza et al., 131–137. Santiago: DIBAM, LOM Ediciones, Universidad Arturo Prat.

Grez, S. 2007. *De la "regeneración del pueblo" a la huelga general: génesis y evolución histórica del movimiento obrero en Chile (1810–1890)*. Santiago: RIL Editores.

Guajardo, G. 1992. El aprendizaje de la tecnología del ferrocarril en Chile, 1850–1950. *Quipu* 9 (1): 17–46.

Guajardo, G. 1993. Tecnología y trabajo en Chile, 1850–1930. *Cuadernos Americanos* 38: 153–179.

Hui, A., L. Roberts, and S. Rockman. 2023. Introduction: Launching a labor history of science. Focus: Let's get to work: Bringing labor history and the history of science together. *Isis* 114(4), 817–826.

King, H.C. 1955. *The history of the telescope*. Mineola, New York: Dover Publications Inc.

Mac, P. E. 1990. Strategies and compromises—Women in astronomy at Harvard College Observatory 1870–1920. *Journal for the History of Astronomy* 21 (1), 65.

Matus, M., ed. 2009. *Hombres de metal. Trabajadores ferroviarios y metalúrgicos chilenos en el Ciclo Salitrero, (1880–1930)*. Santiago: Universidad de Chile.

Nasim, O. W. 2019. The labour of handwork in astronomy: Between drawing and photography in Anton Pannekoek. In *Anton Pannekoek: Ways of viewing science and society*, ed. C. Tai, B. Steen, and J. Dongen, 249–284. Amsterdam: Amsterdam University Press.

Ortega, L. 2005. *Chile en ruta al capitalismo: Cambio, euforia y depresión 1850–1880*. Santiago: Dibam-LOM.

Pinto, J. 1998. *Trabajos y rebeldías en la pampa salitrera: El ciclo del salitre y la reconfiguración de las identidades populares: (1850–1900)*. Santiago: Editorial Universidad de Santiago de Chile.

Salazar, G. 2003. *Historia de la Acumulación Capitalista en Chile*. Santiago: LOM.

Schaffer, S., and S. Shapin. 1985. *Leviathan and the air-pump: Hobbes, boyle, and the experimental life*. Princeton: Princeton University Press.

Schiebinger, L. 1999. *Has feminism changed science?* Harvard: Harvard University Press.

Thoren, V. E., and J. R. Christianson. 1990. *The lord of Uraniborg: A biography of Tycho Brahe*. Cambridge: Cambridge University Press.

Warner, D. J. 1968. *Alvan Clark and Sons artists in optics*. Washington D.C.: Smithsonian Institution Press.

Chapter 3
Buildings and Constructions

Abstract The text examines challenges faced by the National Astronomical Observatory of Chile, including dust accumulation, infrastructure defects, and installation delays. It explores the architectural significance of observatories globally, emphasizing functionality over style and highlighting the fusion of scientific ideals with local practices. The observatory's relocations reveal the intricate labor involved in construction and adaptation, often overshadowed by astronomers' narratives. In this chapter, three locations where the National Astronomical Observatory of Chile was established are analyzed: from the center of Santiago, passing through the western sector, to the south of the city. Using archival materials and accounts from key figures, the chapter examines how the conditions of the city of Santiago, as well as the development of astronomical science itself, rendered the use of certain instruments impractical or hindered sky observations. Attention is given to the role of builders and architects in the dynamics of astronomy, aspects often overlooked when studying the discipline and its advancements in Chile.

Keywords Infrastructure · Astronomical architecture · Architects · Builders

The accumulation of dust on instruments, defective roofs, rainwater seepage, difficulties involving the installation of domes and the effects of vibrations from the nearby railroad are but some of the frequent complaints found in the administrative reports of the National Astronomical Observatory of Chile during the period under study. It's striking how often budgets were approved for repairing infrastructure, addressing a specific problem that soon reappeared as if nothing had been done. And it's equally striking, when reading these reports, to observe how much time went by between the arrival of certain instruments and their final installation due to infrastructure problems.

What's going on here? How can we shed light on the work involved in these problems if, as we shall see, these construction workers, architects and infrastructure repairmen left behind almost no traces? And yet we continue to look for signs of their importance. As mentioned in the introduction, following what Aubin has called the *regime of spatiality*, it is essential to understand astronomy through the installation

© The Author(s), under exclusive license to Springer Nature Switzerland AG 2025   27
C. Sanhueza-Cerda, *The Day Laborers of Science. Technical Work at the Astronomical Observatory of Chile (1852–1927)*, SpringerBriefs in History of Science and Technology, https://doi.org/10.1007/978-3-031-84350-1_3

of instruments in the architecture of buildings, their layout within the observatory and the nature of the geographic and abstract space in which they are found.[1]

The architecture of astronomical observatories is not just a question of styles, although it does reveal a great deal about what they "want to say" to the world. As mentioned in the introduction, observatories tend to be similar to each other, which does not necessarily imply the lack of a local identity, as has been studied in observatories not just in Europe but in colonized spaces as well.[2] Even observatories that seek a local style, such as the one built in Lucknow, India in the 1830s, consider the role of "high technology" dedicated to practical astronomy alongside questions of style. The Lucknow observatory doesn't just represent colonial architecture of European origin, as its design also had to ensure the mobility of the dome, isolate the platform from the vibrations of the rest of the building and optimize its lighting, along with involving climate control.[3] Here we see the fusion of the scientific ideals of both worlds, "mitigated by local practice and contingencies."[4]

It's key to remember the functionality these buildings had to achieve, which allows us to understand the tensions between astronomers, manufacturers and maintenance staff in the quest for stabilization. We know that, during the period under study, astronomy involved the construction of buildings whose placement was crucial, as modern requirements demanded increasingly difficult environments, but also certain astroclimatic conditions that determined the most favorable place for making observations. At the same time, the building complex "must be well calculated, so that the domes should not screen one another, and the South direction should always remain free."[5] This process was not unknown to the National Astronomical Observatory of Chile, whose early years saw it move from the Cerro de Santa Lucía site in downtown Santiago, stopping off at the Quinta Normal in the western zone until it finally established itself at Lo Espejo, in the south of the city. The type of instruments used, as well as the research problems in vogue, led astronomical observatories to be situated "in mountain or sub-mountain areas, or in zones with favorable weather conditions, away from other buildings and luminous sources."[6] These changes, especially since the early nineteenth century, in turn created a new type of architecture for observatories that was characterized by a dome, initially cylindrical or conical and later spherical, "symbolizing the sky, atop of the main building, a feature that is nowadays seen as typical for observatories."[7] Starting in the 1830s, façades with three domes began to emerge in Europe "because of the increased number of instruments being used (refractors, heliometer, meridian circle or transit instrument)."[8] An observatory like the one in Chile, while located in the global periphery, followed in the footsteps

---

[1] Aubin (2017).

[2] See: Dumitrache and Dumitrache (2009, 1–17) and Beuermann (2005).

[3] Bartsch and Scriver (2019, 72).

[4] Bartsch and Scriver. (2019, 73).

[5] Dumitrache and Dumitrache (2009, 12).

[6] Dumitrache and Dumitrache (2009, 12).

[7] Wolfschmidt (2021, 13).

[8] Wolfschmidt (2021, 13).

of its counterparts in Europe, which explains the institution's moves from one part of the city to another.

These moves by the Astronomical Observatory of Santiago, as well as the arguments employed by astronomers to justify them, are the window through which we can catch a glimpse of the labor of those in charge of building and adapting its structures, as well as special components such as domes and pillars. Where should we start? We will review each site chosen for the institution, concentrating on the construction of its infrastructure, as well as the difficulties faced in ensuring the functioning of its instruments due to problems with the facilities and their surrounding environment. As in the rest of this book, we will analyze those setbacks and conflicts that reveal the role of those in charge of ensuring the optimal functioning of the observatory. Sometimes the voices of builders are weak or inaudible under those of the astronomers and evidently attempts to uncover the work of builders and architects is mediated by the available documentation, which is based on official reports written by these same astronomers.

3.1 Cerro de Santa Lucía: From Rocky Hill to Observatory

The history of astronomical observation on Chilean soil began in the mid-nineteenth century, at a time in which observations had to be made across the globe in order to validate the efforts of astronomers. Astronomy was then facing the challenge of coordinating different points of observation to determine the distance between stars and create a measurement system that would allow for the localization of the positions, distances and orbits of celestial bodies. For centuries, different methods had been put forward so that predictable astronomical events could be used for this purpose, which could be studied by calculating the solar "parallax": that is, the angle in the apparent position of an object observed from two different sites. In 1716, Edmund Halley suggested that the "transit of Venus"—which occurs when the planet travels directly between the Sun and the Earth—could be used for this purpose if two observers, situated on more or less the same longitude but with a significant difference in latitude between them, could observe the path of Venus across the Sun at slightly different transects.

Christian Ludwig Gerling (1788–1864) of the Philipps-Universität Marburg in Germany refloated Halley's idea, suggesting that the solar parallax could be determined by measuring the positions of Venus and Mars near their inferior conjunction, especially at stationary points, using observatories situated near the planet's meridian but with a great deal of distance in latitude between them.[9] M. Gilliss (1811–1865), an astronomer at the US Naval Observatory, decided to explore Gerling's ideas by carrying out a series of observations that would later be compared with those made by his colleagues in the US. As the meridian of the East Coast of the United States

---

[9] See: Keenan et al. (1985, 100), Huffman (1991, 208–220), Duerbeck (2003, 3), Dick (2003), Schrimpf (2014) and Hidalgo (2017).

also passes through Chile, Gilliss considered establishing an observation point as far to the south as possible.[10]

But where could this observation site be found? For over six years (before and after the expedition), Gilliss and Gerling exchanged letters discussing the importance of the astronomical expedition to Chile and deliberated on matters such as ways of securing financing, building the necessary observation and measurement instruments and the scientific questions that could be explored.[11] One of the most complex questions was the selection of the most appropriate site for later comparing the measurements made in the Southern and Northern Hemispheres. Where could it be found?

Gilliss had considered setting up an observation point on Chiloé, an island in southern Chile, both because of its geography as well as its inhabitants. While preparing his astronomical expedition to the Southern Hemisphere, Gilliss told Gerling: "Clearly at the same meridian as Washington, but in latitude 43°S, is the island of Chiloe, a place of the same trade which is, I think occasionally visited by American whales ships for supplies, but [in any] event possessing sufficient intercourse with the coast of South America to render it accessible without much trouble and to avoid the necessity of a special ship. I think, inhabited too, by civilized people."[12] Nevertheless, the original idea of reaching more southerly latitudes clashed with his ignorance of the site: the area's rainy climate made astronomical observation a difficult task. Gilliss had not considered that, even in South America, climatic conditions at these latitudes would resemble those of countries in the Northern Hemisphere situated close to the North Pole. Besides, it was impossible for them to install the necessary equipment without urban infrastructure (roads, lodgings, etc.) that would allow them to transport it to the site and guarantee the safety of the research team. In the end, Gilliss had to settle for an observatory close to the Chilean capital. The problem then shifts to the chosen site, to the installation of the instruments and huts they had brought from the United States. Was this simply a matter of importing technology or was local support required to choose a site, build on it and later administer and maintain it?

In Gilliss's report, issued by the US Naval Astronomical Expedition, we learn that the first thing he did was look for a site on the heights surrounding Santiago: "I started for Santa Lucia, the little rocky hill in the eastern portion of the city which had been indicated by the ambassador at Washington as suitable for our purposes."[13] This choice was recommended by the Chilean Consul in the United States, Manuel Carvallo Gómez (1808–1867), who had also intervened so that the country would receive the expedition.[14] Gilliss, continuing his narrative, laments that "the result of the inspection was far from favorable, because of the vicinity of the Andes, of its

[10] Gilliss (1855–1856).

[11] Sanhueza-Cerda and Valderrama (2020).

[12] Letter from Gilliss to Gerling, July 25, 1847, Archive of C. L. Gerling, Library of the Philipps-Universität Marburg, (Marburg).

[13] Gilliss (1855–1856, 453–454).

[14] See: Sanhueza-Cerda and Valderrama (2020, 201).

somewhat precipitous ascent, and the inevitably tedious and expensive labor required to level sufficient space near its rugged summit." The place clearly looked differently in situ than they had imagined it did from the United States. Gilliss sought out other nearby hills, such as Cerro Blanco, that might better serve the expedition's objectives and wouldn't threaten its instruments. Nevertheless, "there were no suitable accommodations for the officers near enough to either of the latter; and as the last was reputed to be excessively wet during winter, it was wholly unfitted for our purposes."[15] So discouraging was the outlook for a site in Santiago that Gilliss made arrangements "to visit Talca as soon as the government approved our establishment in Chile." Nevertheless, Gilliss reconsidered after consulting with "intelligent natives" and "foreign residents" who knew the country well, as they assured him "that while Talca possessed no advantage, so far as distance from the mountains could be considered, it would be impossible to obtain there the facilities which the capital afforded for erecting instruments or their repair in case of necessity."[16] This was crucial: in the end, it wasn't enough to have clear skies and imposing heights. It was a social and political matter, rather than simply a problem of selecting an astronomically and meteorologically apt location. Gilliss himself said as much in his report:

> …the foundation of a permanent observatory on the southern portion of the continent was a great desiderátum, which could only be obtained, they said, by enlisting influential persons in its behalf. This might require the intervention of two distinct classes—scientific men who would appreciate its utility, and political men to vote the necessary outlay. These could nowhere be found so well as at Santiago, about its university and government.[17]

Effectively, as Gilliss said, the role of the government was fundamental in deciding to adapt the Santa Lucía outcrop and transform it into an adequate site for building an observatory. As the North American scientist wrote:

> Government recognised the importance and utility of the work we came to perform, and volunteered every facility within its control, viz: a portion of San Lucia should be levelled for our use, if that hill was selected; rooms in the castle should be placed at our control; a guard should be stationed at the observatories for their and our protection; and everything intended for us should be admitted free of duty.[18]

Once the decision had been made with the support of the government, the latter played a fundamental role in flattening the hilltop, a task assigned to the chief of police, "who had a large gang at work on the tough porphyritic blocks." Given the proximity to the city, explosives could not be used and the "basaltic masses could only be broken down by building fires and suddenly pouring water on the heated rock, or with iron mauls and wedges—both processes necessarily tedious."[19] This work was complemented by the construction of a terrace wide enough "for the smallest

---

[15] Gilliss (1855–1856, 454).

[16] Gilliss (1855–1856, 454).

[17] Gilliss (1855–1856, 454).

[18] Gilliss (1855–1856, 454).

[19] Gilliss (1855–1856, 454).

rotating observatory," for which "it was necessary to build a wall across a short and steep ravine, and fill between the artificial and natural walls with rocks and earth."[20]

The installation of the huts and instruments was likewise difficult, as recounted by Gilliss:

> The first ascent was by a ladder, and thence the vertical height to be overcome is sixty or seventy feet, over an irregular surface of rock, inclined about forty degrees. As the sun was glaringly hot, the labor proved very severe. Stripped as they were, with only pantaloons, hat, and sandals, when the poor fellows deposited their loads beside the excavated trench every muscle of their bodies trembled, and their hearts could be seen throbbing as though they would burst.[21]

This work was done "without incidents." After crossing the continent, this little building "originally built in Washington and then packed up for shipment" was reassembled. Gilliss couldn't hide his contentment: "On the 6th of December I had the satisfaction to obtain a first look through the telescope erected on its pier."[22] We don't know who carried out these tasks, but they were undoubtedly Chilean workers. Here we could make the supposition, based on other experiences during this period, that local construction principles and techniques influenced the building, even if it had been brought here from elsewhere.[23] Who made repairs when needed? Were there adaptations? Gilliss has nothing to say on this point.[24]

As he finished recounting the achievement of adapting Cerro de Santa Lucía for astronomical observation, Gilliss didn't conceal the role that this undertaking would play for Chilean science: "Santiago through our influence established the first national observatory of South America."[25] And so it would be: following the departure of the US Naval Astronomical Expedition in 1852, a bilateral contract was signed allowing the Chilean government to acquire these instruments and buildings. The National Astronomical Observatory of Chile was born.

The astronomer Karl Moesta was named the first director of the Chilean observatory. According to the institution's official communications, especially its reports and letters to the Chilean Ministry of Public Instruction, which administered the observatory, he gave continuity to the work and concerns of the US expedition. Nevertheless, this continuity depended on the material conditions through which observations were made. From the start, this observatory had been designed to be temporary and its huts were installed on the hill without pillars that were sufficiently sunk into the rock, nor were excavations performed to deepen the terrain and thus help stabilize its instruments.[26] When performing measurements, Moesta soon realized that the

---

[20] Gilliss (1855–1856, 454).

[21] Gilliss (1855–1856, 454).

[22] Gilliss (1855–1856, 454).

[23] Bartsch and Scriver (2019, 59–77).

[24] Regarding the placement of the Gilliss expedition's observatory at Cerro Santo Lucía, see: Saavedra (2018, 152–163).

[25] Gilliss (1855–1856, 454).

[26] Keenan et al. (1985, 108).

azimuth of the transit instrument shifted periodically, especially during the night. What could be causing this problem?

The astronomer found an explanation in the geological foundations upon which the observatory had been built. After dismissing movements in the instruments due to seismic activity, Moesta verified their profound oscillations from day to night by making daily meteorological observations. It then followed that the instruments had been affected by "the dilation and contraction of the porphyritic columns which support the stone that serves as a foundation."[27] Was it viable to continue conducting observation work at this site? For Moesta, "it can easily be concluded that, in the Santiago area, there is perhaps no other place less suitable for an observatory than Cerro de Santa Lucía, it being of the utmost importance to establish a site where atmospheric influences have the least possible effect on the position of the instruments."[28] It had become urgent to move the observatory "far from the city center."[29]

In his 1857 annual report, Moesta told the Minister of Public Instruction about the difficulties created by the infrastructure of the Santa Lucía observatory, writing that, "despite the repairs made to the three huts in which the observatory's instruments had been installed" they were "so insecure, due to the poor condition of the wood of which they were made, that one of them was entirely destroyed in the storm on June 15th." He then mentioned that the second equatorial telescope "came apart (…) and its precious eyepiece was smashed, leaving the rest of the instrument useless." The other huts had been "repaired on repeated occasions and I hope, with the last repair performed quite recently, that they will endure until the new site is finished."[30] His report addressed the possibility of having a new building and better working conditions even more directly. For Moesta, such a decision would open up "a new era for the Observatory of Chile (…) through the government's willingness to build a solid, adequate building for an astronomical observatory at the Quinta Normal, thus allowing this establishment not only to find itself in conditions to provide its work with that exactitude required by science today, but making its existence forever secure at the same time."[31]

The so-called Quinta Normal de Agricultura, built years beforehand as an exhibition and acclimatization center for plants and livestock, raised the hopes of astronomers that they would have an observation site where they could control those factors that destabilized their precision instruments.[32] This *place of science*, which would later host the Museum of Natural History, was held up as an ideal site for astronomy due to its distance from the city and its urban bustle, not to mention the dust clouds raised by the movements of carriages, which also affected the telescopes.

---

[27] Moesta (1853, 61–64).

[28] Moesta (1853, 64).

[29] Keenan et al. (1985, 108).

[30] Archivo Nacional de Chile, Fondo Ministerio de Justicia e Instrucción Pública, vol. 84, May 20, 1857.

[31] Archivo Nacional de Chile, Fondo Ministerio de Justicia e Instrucción Pública, vol. 84, May 20, 1857.

[32] Regarding the Quinta Normal de Agricultura, see: Hecht (2016) and Montealegre (2017).

We know the reasons for this move, but very little about those who carried it out. Can we explore the world of its builders, foremen and architects through the design of the new buildings at the Quinta Normal?

3.2 La Quinta Normal de Agricultura: A Place for Science

Despite Moesta's hopes of having a new building and aspirations of ensuring the continuity of his observations, the construction and move were slow and complicated. This meant that the dismantlement of certain pieces and telescopes in order to be taken to the new site wasn't always coordinated with construction at the Quinta Normal. Moesta informed the Minister of Public Instruction of his concern for the deterioration of the instruments that were still at Cerro de Santa Lucía. The astronomer was likewise concerned with the most sensitive aspect of the new infrastructure: "the construction of domes and pillars, as well as (…) other minor work on the interior."[33] The building's structure, zinc roof, doors, windows and stairs weren't particularly relevant to the stabilization of its observations in comparison to the domes and the pillars upon which the instruments were to be installed. In Moesta's eyes, it was particularly worrying that the corresponding plans were not even ready: it was "an absolute necessity that the architect of this great work draw up such plans in accordance with the conditions of the establishment's instruments." Moesta wrote to the government authorities that "details are of great importance for the observatory complex, in accordance with the order required by the prudent construction (of the) site."[34] Are what Moesta called "details" anything but the heart of the observatory? They were the conditions that would ensure the precision of its astronomical work. Here we can see that infrastructure does not just refer to the heavy work, but above all to the stabilization of the instruments. Moesta complained to the minister that construction work had not been coordinated in accordance with the requirements that would allow the instruments to function correctly: "On this point, it's important for me to say that the introduction and placement of the large solid stones in the building's wings that are to be used as the pillars for the meridian instruments will be quite difficult now that the walls of these sections have already been concluded." Later on, he grumbled about the secondary "little tower" and its "rather unfortunate placement that will leave the piece for the primary vertical somewhat useless."[35] Thanks to the available documentation, we know that the Civil Corps Office was in charge of construction and that Moesta never reviewed their plans.[36]

---

[33] Archivo Nacional de Chile, Fondo Ministerio de Justicia e Instrucción Pública, vol. 84, May 26, 1859.

[34] Archivo Nacional de Chile, Fondo Ministerio de Justicia e Instrucción Pública, vol. 84, May 26, 1859.

[35] Archivo Nacional de Chile, Fondo Ministerio de Justicia e Instrucción Pública, vol. 84, May 26, 1859.

[36] In the National Archive of Chile, there are drawings of the buildings which could be considered to be plans, but they are very imprecise.

In the following year's annual report, the situation of the National Observatory seemed to have improved, as Moesta had become more involved in the construction process and the movement of instruments. Once again, the report centered on what could be called the building's accessory elements rather than on the heavy work. Moesta recounted the problem with the hall "designated for the meridian circle" due to the complex work "in the foundations for the instrument's pillars and the pendulum." The astronomer continued: "What caused the most work was finding solid rocks for the pillars upon which the meridian circle and the pendulum are to be suspended, as it's difficult to find large pieces of rock at the quarries of Cerro San Cristóbal that do not have fissures."[37] These reports reveal the materiality required and the control that had to be exercised over the entire process. Moesta was ultimately satisfied with the results, as he mentioned to the minister: "All three pillars were introduced into the hall and installed on their respective foundations with felicity and skill. To increase the stability of these pillars they have been given three feet at their base and I have every reason to believe that the installation of the meridian circle upon these pillars will be carried out satisfactorily. The aforementioned hall will soon be ready to receive said instrument, with the sole exception of the two iron levers through which the meridian openings in the roof are to be closed."[38]

Who had done all this work? As we have seen, Moesta mentioned that the Corps of Civil Engineers was in charge of construction, but we know nothing about their workers. Yet when reviewing the archives, we can see that Moesta mentions that the metal pieces, such as the levers and the building's central dome, "were assigned to the School of Arts and Trades." This school was an educational institution for training technicians and specialized laborers and, according to the available documentation, was in charge of fabricating those pieces, which were built separately from the heavy work.[39] It could be said that these pieces were minor aspects compared to the construction of the buildings themselves, but they were crucial when installing telescopes. A large part of Moesta's subsequent reports even refer to his problems with the School of Arts and Trades for not having readied the components needed to move the domes' doors, insisting that "it's necessary for me to inform you that neither the aforementioned levers nor the simple mechanism needed for moving them have been finished as of yet and, even if they had been finished, it does not seem that they offer the ease of use that would have been desired. The lack of said levers is the only reason for which the meridian circle has not yet been installed in its corresponding place." This problem was so serious that the 1860 report ended by stating that "the

[37] Archivo Nacional de Chile, Fondo Ministerio de Justicia e Instrucción Pública, vol. 84, May 26, 1859.

[38] But not everything revolved around domes and the installation process. There was also the problem of airborne dust, which had already affected the observatory at Cerro de Santa Lucía. Moesta suggested a possible solution to the minister: "As it is very important to protect the observatory's instruments from the effects of dust to the extent that it is possible, at the current site there's no better manner of achieving this objective than planting sufficient shrubbery on the grounds which the Supreme Government has conceded to the institution." See: Archivo Nacional de Chile, Fondo Ministerio de Justicia e Instrucción Pública, vol. 84, May 26, 1859.

[39] See: Castillo (2016).

conclusion of the new observatory now principally depends on the conclusion of these two domes."[40] Astronomical observations had been interrupted by the lack of available pieces, as the astronomer made clear in his 1861 annual report to the Minister of Public Instruction: "Of the two wings of the building built for the meridian instrument, one has been ready to receive the Great Meridian Circle since the month of February; nevertheless, as the School of Arts and Trades has taken an extraordinary time in finishing the two simple levers needed to open and close the openings in the roof, it was not until the month of May that the hall had been definitively prepared and we could proceed to install the aforementioned device. The other wing has remained unusable for the entire year due to the same reason."[41] According to David Edgerton, the historiography of technicians and of technique-society relations has primordially centered on innovation, which it fails to distinguish "from the study of technology in widespread use, which is necessary old, and is often seen as out-of-date, obsolete, and merely persisting."[42] These reflections lead us to a very different vision, one found by paying attention to the technology in use. It is precisely this shift in perspective that reveals the role of other actors in the construction of the observatory beyond the architects and engineers. What others ignore is what allows us to draw these other actors out of their silence.

Following the departure of Karl Moesta, Juan Ignacio Vergara (1837–1889) became the director of the National Astronomical Observatory of Chile. In his first report from 1865, he informed the minister that the building was functioning correctly "with the exception of its roof, which, being made of zinc with its sheets hammered in every which way, frequently breaks due to the effects of dilation. This obliges us to make repairs each winter to keep rainwater from entering." For the new director, this was clearly a matter of maintenance or "light repairs."[43] As with the construction of the buildings, we come back to the issue of the technologies in use, which includes the maintenance process (as we have already seen with the instruments) but also the difficulties faced by astronomers when the infrastructure didn't fulfil the purpose of protecting their telescopes and allowing them to function. The documentation studied herein offers a periodic record of these problems and difficulties with the buildings. In the following year's report, Vergara informed the minister that, despite having inspected the leaky roof, they had been unable to repair it: "Unfortunately, this work cannot be considered to have been finished as long as the roofing material has not been completely replaced (...) the zinc that roofs the building constantly (breaks), such that it is impossible to avoid the entrance of rainwater in winter."[44] The problem of maintenance, which could already be seen with the builders and

[40] Archivo Nacional de Chile, Fondo Ministerio de Justicia e Instrucción Pública, vol. 84, June 14, 1860.

[41] Archivo Nacional de Chile, Fondo Ministerio de Justicia e Instrucción Pública, vol. 84, May 26, 1861.

[42] Edgerton (1999, 112).

[43] Archivo Nacional de Chile, Fondo Ministerio de Justicia e Instrucción Pública, vol. 148, May 15, 1865.

[44] Archivo Nacional de Chile, Fondo Ministerio de Justicia e Instrucción Pública, vol. 148, April 18, 1866.

workers, emerges here as a budgetary issue, whether when requesting funding or complaining about problems that have not been resolved. The documentation does not offer us any information on the actions taken by those in charge.

One issue that appears in Vergara's reports to the minister was that, beyond pressing issues of infrastructure maintenance and repair, the very placement of the new facilities at the Quinta Normal had become problematic, to the point that it threatened the proper functioning of its instruments and observations. Only ten years had gone by since the observatory's move and its director had already begun to complain to the minister about problems that he said threatened "the conservation of the instruments." The astronomer argued that this was "a result, above all, of the neglect and consequent poor state of nearly all the roads at the Quinta Nacional de Agricultura, principally those surrounding the observatory." He blamed "the heavy traffic of the Quinta's own animals on these roads," which he said meant "that, in the summers, immense clouds of dust rise up in the surroundings of the establishment at all times, and which are deposited on even the most delicate and hidden instruments."[45] In his report, Vergara informed the minister of "the magnitude of the resulting damage if an immediate remedy is not applied." He then explained the paradoxical situation caused by the observatory's move: "It would be enough to observe that one of the primary reasons that was taken into consideration when moving the observatory from its old site at Cerro de Santa Lucía to the one it occupies now, outside the city, was the dust that is inevitably raised by any settlement; this is also one of the reasons why, in all corners of the world, astronomical observatories tend to be located far from the most populous neighborhoods in those cases in which it is impossible to remove them from the cities completely." His conclusion was clear: "If the problem that I have just described were to continue indefinitely, after moving the observatory to a new site at considerable cost, we would obtain, along with the contrary result to the one pursued, the disablement of the instruments."[46]

As we have seen, changes within the discipline were making it increasingly difficult for observatories to remain in cities. Despite the need for a move, the economic difficulties involved in making it a reality forced Vergara to turn to other options, as revealed in the proposal he made to the minister in March 1875. As the international exposition that was to be held at the Quinta Normal required the movement of great quantities of earth, Vergara suggested taking advantage of this process to isolate the observatory. Vergara's arguments are interesting because they reveal the problems with finding a home for the institution. The astronomer thus described the observatory's situation to the minister:

> Ever since traffic began on the northern railroad, whose many heavy trains pass less than two hundred meters from the observatory, I have noticed irregular alterations in the positions of the instruments that require me to execute frequent corrections, prejudicial to their

---

[45] Archivo Nacional de Chile, Fondo Ministerio de Justicia e Instrucción Pública, vol. 148, May 29, 1869.

[46] Archivo Nacional de Chile, Fondo Ministerio de Justicia e Instrucción Pública, vol. 148, May 29, 1869.

conservation, in order to avoid the interference of these alterations in the work for which these instruments have been designed.[47]

Clearly, the advantages of a site located far from the city center had been lost in a short period of time due to the vigorous development of Santiago's railways. Vergara said that, when the observatory was built, "no one knew that, in a very near time, a railway would perhaps pass by at such a short distance as the one I have indicated, and therefore didn't take the trouble of isolating the foundations on which the instruments rest." This situation meant that "the vibrations that the trains produce in the ground are constantly transmitted to the instruments, and these vibrations are the cause of the changes in position experienced by the instruments from day to day."[48] It would have been ideal to move the observatory again, but Vergara, being practical, proposed something that was more efficient and imaginative: given that the observatory "has been definitively established, we are only faced with the possibility of isolating the entire institution with a surrounding trench whose depth would be sufficient for intercepting those vibrations. This trench could be left empty or filled with water or sand, without preference."[49] Who was to pay for this work and what was to be done with this enormous volume of earth? Vergara explained his plan: "As the land that (...) requires an embankment (...) for the exposition is situated next to the observatory, the construction of the trench will currently have the double advantage of avoiding, at no cost, the aforementioned problems facing this institution, and of providing cheap and abundant material for these embankments."[50]

No information has survived regarding who built the observatory at the Quinta Normal. How can we confront this silence? As in other cases, we must seek the profile of the builders in the work itself. Perhaps we lack the names and responsibilities of those who built the buildings, but we do know who made the pillars on which the telescopes were installed. While this doesn't concern the infrastructure itself, the creation of artifacts associated with the proper installation of instruments was crucial. Vergara reported on the creation of the pillars for the meridian circle that had recently been acquired by the observatory. On a set of expense reports dated August 16, 1870, the name of Richard Brown, "Government Architect" appears.[51] On other documents, one can note his participation in the construction of towers for other instruments, as occurred with the installation of the equatorial telescope in 1874. It's all we know: his name and position.

---

[47] Archivo Nacional de Chile, Fondo Ministerio de Justicia e Instrucción Pública, vol. 148, March 27, 1874.

[48] Archivo Nacional de Chile, Fondo Ministerio de Justicia e Instrucción Pública, vol. 148, March 27, 1874.

[49] Archivo Nacional de Chile, Fondo Ministerio de Justicia e Instrucción Pública, vol. 148, March 27, 1874.

[50] Archivo Nacional de Chile, Fondo Ministerio de Justicia e Instrucción Pública, vol. 148, March 27, 1874.

[51] Archivo Nacional de Chile, Fondo ministerio de Justicia e Instrucción Pública, vol. 148, August 16, 1874.

What was the fate of this place for science? The documentation reveals that, by the end of the nineteenth century, the observation space at the Quinta Normal was reaching its limits. Besides the contingencies faced by Vergara or his new ideas for stabilizing observations, it was evident that the National Astronomical Observatory could not continue to be located at that site. Something had to be done, but creating a new observatory required a political context and official support that Vergara lacked. It would not become a reality until the dawn of the next century.

3.3 Lo Espejo: A Site for Astronomy

By the end of the nineteenth century, it had become clear that the National Astronomical Observatory of Chile was suffering from the problems caused by its location, given the city's expansion and the corresponding light pollution, which threatened the continuity of its work and even the very existence of its instruments. By this time, José Ignacio Vergara was no longer very involved in the institution's administration. He was also serving as the mayor of Talca as well as the federal Minister of the Interior, both very demanding roles. Official reports reveal that the director was increasingly being substituted in his duties, which could have affected its level of government support. Complaints regarding the state of its buildings were recurrent in its communications to the Ministry of Justice and Public Instruction, which administered the observatory.

As at other government institutions, the absence of the director clearly made funding more complicated. Even the acquisition of new instruments was jeopardized by the lack of proper facilities, as can be seen with the Great Equatorial Telescope that had been ordered from Europe. Substitute Director Ruperto Solar informed the minister that they lacked "the necessary funds to put a provisional covering over the tower that had been left unfinished in 1873, built with the objective of mounting the Great Equatorial which the establishment has possessed since that date." He drew a clear conclusion: "This building, exposed as it is to the ravages of time, is in danger of completely deteriorating. The instrument that it was to receive thus remains in storage, and I suppose in good condition, as the boxes that contain it have never been opened."[52]

We know from the official documentation that the architect Ricardo Brown designed the building where the Great Equatorial Telescope was to be installed in 1879 but that it wasn't finished until 1882. Here we see not only problems with the budget, but also with the dome itself, which couldn't be coordinated with the movement that the instrument required. The difficulty lay in the rotation mechanism, which had been assigned to the mechanic Desiderio Corbeana but had not yet been built. The instrument could not be used without a mobile dome, even though the building housing it was ready. Here we can see that builders and architects were not

---

[52] Archivo Nacional de Chile, Fondo Ministerio de Justicia e Instrucción Pública, vol. 148, June 2, 1879.

enough. José Ignacio Vergara had to directly intervene so that the necessary pieces could be finished and the Great Equatorial Telescope thus made operational. Vergara worriedly informed the minister of these delays in construction, writing that "men of science lack knowledge of such things." He wrote that, while the carpentry work on the inside of the tower had not been finished, what remained to be done was not that significant: "If the machinery destined to produce the rotating movement is added soon, as I was told it would be, as it has supposedly already been finished in the mechanics' workshop, there wouldn't be (…) in fact any serious inconvenience."[53] This case reveals that the structures and operations that made telescopes function correctly didn't fall under the purview of the astronomers (who, in fact, had to trust the builders). It can also be seen here that work on observatory infrastructure didn't just involve erecting buildings but ensuring that their instruments functioned properly. The observatory depended as much on its layout, height and the orientation and mobility of its dome as it did on the ability of its astronomers, observers and collaborators.

As we have seen, the technical skill or function of builders was not enough to stabilize observations. At the end of the nineteenth century, despite the efforts of astronomers, collaborators and technicians, the observatory's placement at the Quinta Normal continued to represent an obstacle. The trains were not the only problem, as there was also the dust from carriages and the movements of the animals at the Zoological Garden that had been opened at the park. Nevertheless, Vergara was unable to secure the observatory's move; without political support, the astronomical institution was reduced to a meteorological observatory. Vergara died in 1899. The institution was further weakened and distanced from the international astronomical projects that had given it a role in the circuit of observatories. Its instruments were deteriorating and it barely had the minimum staff to keep up observations.[54]

Chance, that elusive factor in history, changed everything at the dawn of the twentieth century. When Pedro Montt became president in 1906, the political outlook shifted, as did the level of support for Chilean science. It seems that Montt took a special interest in astronomy and understood the need for funding and new technical staff. The institution had a new opportunity to demonstrate that these activities—so far removed from the concerns of the masses—could be performed in Chile. But how could the observatory be made functional without trained staff? As new domestic astronomers had not been trained for some time, Pedro Montt followed the example of his father, President Manuel Montt from the mid-nineteenth century, and recruited German scientists to give a new impulse to this southern observatory. The most important was its new director Friedrich W. Ristenpart (1868–1913), who was accompanied by Walter Zurhellen (1880–1916) and Richard Prager (1883–1945), who respectively served as directors of the astrophotography and calculations departments.[55] This new staff was essential for the institution's insertion into global astronomical projects such

53 Archivo Nacional de Chile, Fondo Ministerio de Justicia e Instrucción Pública, vol. 148, June 2, 1879.
54 Keenan et al. (1985, 128).
55 Keenan et al. (1985, 129).

as the Carte du Ciel, which had been resumed with additional funding and support from the Chilean state.[56]

The first thing the German astronomers did in this rebirth of the observatory was address the problem of the site the institution occupied at the Quinta Normal. In 1909, Ristenpart began to design a comprehensive plan for a new observatory that would involve the institution's move to the area south of the city known as Lo Espejo. In his first annual report to the Minister of Justice and Public Instruction, Ristenpart recounted how, after reviewing alternatives to the east of Santiago, he decided on a site located to the south as he believed that the city would expand to the west, which could cause "troubles" for the new observatory. The site he chose clearly offered greater advantages "as it is situated on high ground, with the surrounding lowlands completely clear." This location to the south also made it possible to have an unobstructed view in every direction, as there were neither buildings nor human traffic, but above all because "in no direction is the horizon narrowed by more than five degrees."[57]

Ristenpart not only selected the site, but also designed an entire astronomical complex with "a main building crowned by two domes for the smaller telescopes, three domes arranged for refractors, two buildings for the meridian circles, situated on the same meridian and free of smaller structures for portable instruments, a house for the director, seven rooms for married astronomers and a house for bachelors (…) facilities for the machine room, etc." (see Fig. 3.1).[58]

In this same report, Ristenpart detailed the work that had to be done to prepare the grounds so that they could be used as an observation site. The director put the astronomer Alfredo Weber in charge of "leveling and flattening the terrain; diverting the course of the water" and drawing up plans for the installation of the astrophotographic telescope, as well as "constantly supervising the construction of the buildings."[59] We don't know the workers who carried this out and, in his reports to the minister, Ristenpart only mentioned those governmental authorities who had supported the new observatory, from the president to cabinet ministers and subsecretaries.

---

[56] The Carte du Ciel was a network that sought to capitalize on the new techniques and instruments developed by the Henry brothers to photograph the stars for astronomical ends. In 1887, 20 observatories committed to collectively photograph the entire night sky, seeking out stars at predetermined magnitudes. One of these observatories was the National Astronomical Observatory, which was put in charge of photographing the sky between the declinations of 17° and 23°. The relevance of Chile's participation not only lay in its receiving photographic plates and a Gautier telescope from Paris, but above all in its participation in the development of new, emerging scientific techniques that were modeled by international practices. On the Carte du Ciel, see: Aubin (2003) and Bigg (2000, 106). References taken from Nasim (2017, 181). Regarding the Chilean role, see: Keenan et al. (1985, 124).

[57] Ristenpart (1910, 754). This has been partially analyzed in Sanhueza-Cerda (2022).

[58] Ristenpart (1910, 754).

[59] Ristenpart (1910, 754).

Vista perspectiva del nuevo Observatorio

Fig. 3.1 Photo taken from the Legacy of the Ministry of Public Works of Chile in the National Archive of Chile, "Public Works" section, 1911

One interesting aspect of this new site for science was the fact that the buildings were built specially for the functioning of the instruments—and not just for the observatory's existing telescopes, but also recent acquisitions that had not yet been installed and even future purchases. The wide-open horizon at Lo Espejo allowed for the site to be designed around its observation instruments for the first time. Reviewing the plans for the new site at Lo Espejo in the archives of Chile's Ministry of Public Works, it becomes clear that the instruments were neatly arranged amongst themselves in order to function better together (see Fig. 3.2). Here we can see how the Grubb equatorial telescope (the largest and, therefore, in greatest need of inclination) was located in the center, arranged latitudinally. The two meridian circles were installed in a north–south direction. To the south of the Grubb equatorial telescope was an Eichens equatorial telescope and an astrophotographic equatorial telescope. They were all arranged so as to allow observation without interventions or interference. It was essential that no instrument affect the horizon of visibility.[60]

Ristenpart's plan included the acquisition of a new meridian circle, which was to function alongside the one the observatory already possessed. We have the 1908–1913 correspondence between Ristenpart and the German manufacturer, Repsold and Söhne of Hamburg, in which they discussed critical aspects of the instrument's installation in Chile. The most sensitive points had to do with technical staff and the

---

[60] See: Ministerio de Obras Públicas, Dirección de Arquitectura. Image taken from its website on May 4, 2020: http://www.afda.cl/resultado.php?busq=Observatorio+Astronomico&pag=1& modo=0&v=8d4ec65926fee7ac5438f39bb6faa6d36327ab98b4eb035fda2591497ce5c23c4553b aeeaaf764e8e35de5ed437f2f616d5889797cdb6fa3fb1f57133f6d856f. Retrieved on July 9, 2020.

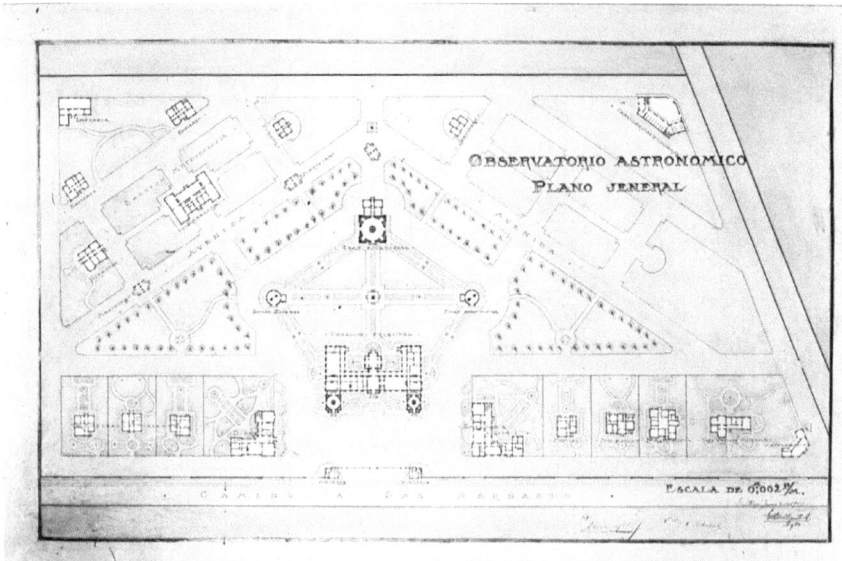

Fig. 3.2 Photo taken from the Legacy of the Ministry of Public Works of Chile in the National Archive of Chile, "Public Works" section, 1911

auxiliary equipment needed for its calibration, as well as the physical location where the new telescope was to operate.[61]

Simultaneously preparing the buildings and telescopes allowed for better coordination and more efficient placement. Unlike at the observatory's other sites, such as Cerro de Santa Lucía or the Quinta Normal, the instruments were able to mutually reinforce each other. Ristenpart mentioned this to the Hamburg manufacturer: the idea was that (as can be seen in Fig. 3.1) both the old and the new meridian circle would be located on the same meridian, so that they could be collimated.[62] Ristenpart thus decided to situate them in a "semicylindrical structure in the upper part of the building."[63]

This coordinated arrangement of buildings and telescopes also influenced decisions on the dimensions of the hut and the dome for the new meridian circle. Ristenpart didn't have any issues with the size of the instrument itself, as he mentioned to its manufacturer: "At the end of your letter, you asked about the dimensions of the room in which the meridian circle will be housed, but as the new observatory is still under construction, I'm lucky enough to be able to adapt the building to the dimensions of your instrument."[64] We can conclude from their correspondence that

---

[61] See: Sanhueza-Cerda (2022).

[62] Collimation is an operation through which a telescope's orientation is adjusted in order to obtain parallel rays of light using a focus. This allows for the precise orientation of the optics within a telescope. See: Hodam 1967).

[63] Letter from Ristenpart to Repsold, February 21, 1909, Hamburg State Archive, A II 28.

[64] Letter from Ristenpart to Repsold, February 21, 1909, Hamburg State Archive, A II 28.

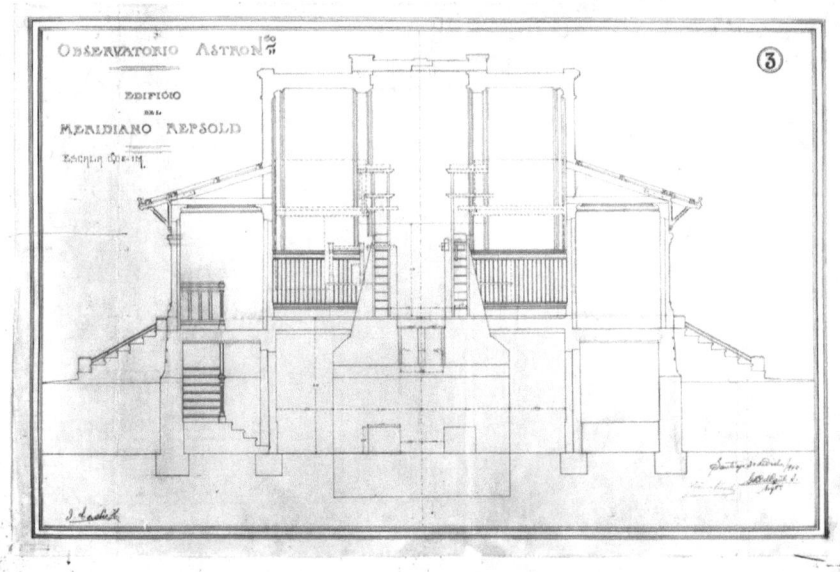

Fig. 3.3 Photo taken from the Legacy of the Ministry of Public Works of Chile in the National Archive of Chile, "Public Works" section, 1911

these two men discussed the design of the space that would house the meridian circle. In a letter dated June 11, 1910, Ristenpart commented that "finally I can begin on the design and construction of the meridian house in accordance with your outline, and I hope to finish it before the meridian circle arrives."[65] From the plans for the meridian circle building at Lo Espejo, it can be seen how it was designed to allow for the ascending movement of the meridian circle, as well as that this dome was the only one that allowed for the entrance of light to be controlled (see Figs. 3.3 and 3.4).

One very important aspect of the instrument's placement had to do with the light within the dome. Ristenpart carried out many lighting tests before the new instrument arrived, as he explained to Repsold: "I've tried many different windows in our observatory, but neither the tallest nor the most beautiful have given me enough light on the white surfaces (that is, on all of them at the same time)." A series of windows oriented toward the zenith had to be built so that "enough light could enter in a staggered fashion." Ristenpart opted for a skylight because, as he mentioned to the manufacturer, "lighting with a skylight is the best for our instrument, which is why we have also built measurement rooms with skylights in our observatory."[66]

The procedures used, the installation instructions included with the sale of the instrument and international protocols (applied here for dome design, lighting and

---

[65] Letter from Ristenpart to Repsold, June 11, 1910, Hamburg State Archive, A II 28.

[66] Letter from Ristenpart to Repsold, April 18, 1910, Hamburg State Archive, A II 28. See Fig. 3.3.

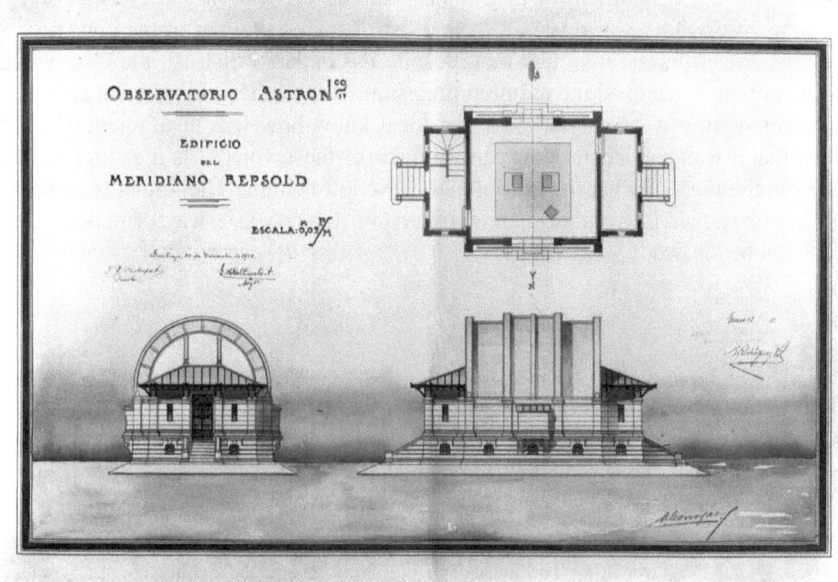

Fig. 3.4 Photo taken from the Legacy of the Ministry of Public Works of Chile in the National Archive of Chile, "Public Works" section, 1911

coordination with other instruments) were not enough to guarantee the success of this endeavor. On June 14, 1911, one month after the instrument's arrival in Chile, Ristenpart was still unable to conclude its installation as he was unable to understand the mechanism through which the base would be affixed to the pillars that anchored the meridian circle to the ground. Was he doing something wrong? Ristenpart told Repsold that he couldn't find "the holes that permitted such an adjustment." His letter ended with a request: "Perhaps you could be so kind as to enlighten us on this point."[67] The stabilization of the observation site was undoubtedly mediated by the local. Upon whom could they rely if the instrument failed to function because of its improper installation? Ristenpart told the manufacturer that their "great distance" and "the country's isolation" from technological development meant he could rely on "no one but myself."[68] No matter how closely they followed international protocols or consulted with the manufacturers, it was ultimately the local that ensured the global movement of a precision instrument.

We are unable to entirely understand Ristenpart's distrust of local labor, as revealed in the available documentary sources. Was there not sufficient technical know-how in Chile? Did Ristenpart fail to recognize the role played by Chilean workers, architects and builders? The correspondence between the manufacturer and the end user, as well as the director's reports to the government minister, do not reveal the other actors

[67] Letter from Ristenpart to Repsold, June 14, 1911, Hamburg State Archive, A II 28.

[68] Letter from Ristenpart to Repsold, June 14, 1911, Hamburg State Archive, A II 28.

involved in the installation of instruments, nor in the design and construction of the new observatory. At most, we have a list of materials used in the construction process. Nevertheless, it's clear that, despite Ristenpart's distrust, the observatory would not have achieved the required precision without the work of local architects and construction workers. The fact that local know-how was insufficient does not mean that it was inexistent. How can we uncover these voices? Is it enough to read between the lines of what the astronomers have left behind? The title of this chapter refers to *constructions* rather than *construction workers* as a way of making clear what can be known.

References

Aubin, D. 2003. The fading star of the paris observatory in the nineteenth century: Astronomer's urban culture and observations. *Osiris* 18: 79–100.

Aubin, D. 2017. L'observatoire: Régimes de spatialité et délocalisation du savoir, 1769–1917. In *Histoire des sciences et des savoirs de la Renaissance à nos jours*, ed. Dominique Pestre, vol. 2, 54–71. Paris: Le Seuil.

Bartsch, K., and P. Scriver. 2019. The house of stars: Astronomy and the architecture of new science in early modern Lucknow (1831–49). In *Ilm. Science, religion and art in Islam*, ed. S. Akkach, 59–78. Adelaide: University of Adelaide Press.

Beuermann, K., ed. 2005. *Grundsätze über die Anlage neuer Sternwarten Georg Heinrich Borheck*. Göttingen: Universitätsverlag.

Bigg, C. 2000. Photography and the labour history of astronomy: The Carte du Ciel. *Acta Historica Astronomiae* 9: 90–106.

Castillo, E. 2016. The school of arts and trades in Sabtiago (EAO), 1849.1977. *Desing Issues* 32 (1), 32–40.

Dick, S. J. 2003. *Sky and ocean joined. The U.S. Naval observatory 1830–2000*. Cambridge: Cambridge University Press.

Duerbeck, H. W.: National and international astronomical activities in Chile, 1849–2002. In *Interplay of periodic, cyclic and stochastic variability in selected areas of the H-R diagram*, ed. C. Sterken, 3–20. San Francisco: San Francisco Astronomical Society of the Pacific (2003).

Dumitrache, C., and D. Dumitrache. 2009. Architectural evolution of astronomical observatories. *Romanian Astronomical Journal* 19 (1): 1–17.

Edgerton, D. 1999. From innovation to use: Ten eclectic theses on the historiography of technology. *History and Technology: An International Journal* 16 (2): 111–136.

Gilliss, J. M. (1855–1856). *The United States astronomical expedition to the Southern Hemispheres in 1849–'52*, vol. 2. Washington: Nicholson Printer.

Hecht, R. 2016. Dissecting the origins of Chile's Quinta normal de agricultura as a Colonial Garden, 1838–1856. *Studies in the History of Gardens & Designed Landscapes* 37 (4): 273–293.

Hidalgo, G. 2017. Revisiting J. M. Gilliss' astronomical expedition to Chile in 1849–1852. *Journal of Astronomical History and Heritage* 20 (2), 161–176.

Hodam, F. 1967. *Technische optik*. Berlin: VEB Verlag Technik.

Huffman, W. 1991. The United states astronomical expedition (1849–52) for the solar parallax. *Journal the History of Astronomy* XXII, 208–220.

Keenan, P.C., S. Pinto, and H. Alvarez. 1985. *El Observatorio Astronómico Nacional de Chile (1852–1965)*, 1985. Santiago: Universidad de Chile.

Moesta, K. 1853. Geología. Observación de un notable fenómeno, que presenta el cerro de Santa Lucía. *Anales de la Universidad de Chile*, 61–64.

Montealegre, P. 2017. *La figuración de un jardín público: Urbanismo y agricultura en la construcción del Santiago moderno (1838–1875)*. Doctoral thesis. Pontificia Universidad Católica de Chile.

Nasim, O. 2017. Observatorium. In *Handbuch für Wissenschatsgeschichte*, ed. Marianne Sommer, 180–192. Stuttgart: J. B. Metzler Verlag.

Ristenpart, F. 1910. El Observatorio Astronómico Nacional de Santiago en 1909. *Anales De La Universidad De Chile* 127: 737–757.

Saavedra, C. 2018. Las terrazas astronómicas del cerro Santa Lucía: Emplazamiento y vestigios del primer observatorio astronómico en Chile, por parte de la expedición astronómica norteamericana de James Gilliss en 1849 1852. In *Intersecciones 2018. III Congreso Interdisciplinario de Investigación en Arquitectura, Diseño, Ciudad y Territorio*, 152–163.

Sanhueza-Cerda, C. 2022. Stabilizing local knowledge: The installation of a Meridian circle at the national astronomical observatory of Chile (1908–1913). *Isis* 113 (4): 710–727.

Sanhueza-Cerda, C., and L. B. Valderrama. 2020. Finding a point of observation in the global south: The C. L. Gerling and J. M. Gilliss correspondence (1847–1856). *Journal for the History of Astronomy* 51 (2), 187–208.

Schrimpf, A. 2014. An international campaign of the 19th century to determine the solar parallax: The US Naval expedition to the southern hemisphere 1849–1852. *European Physical Journal H* 39 (225): 225–244.

Wolfschmidt, G. 2021. Cultural heritage of observatories in the context with the IAU-UNESCO initiative—Highlights in the development of architecture. In *Advancing cultural astronomy: Studies in Honour of Clive Ruggles*, ed. E. Boutsikas, S. McCluskey, and J. Steele, 291–314. New York, Berlin, Heidelberg: Springer.

Chapter 4
Mechanics

Abstract This chapter delves into the pivotal role played by technicians at the National Astronomical Observatory in dismantling, maintaining, and repairing instruments. Its aim is to underscore the convergence of materiality and the consolidation of knowledge, shedding light on the significance of their contributions, which have hitherto been neglected. By delving into their biographies, this study uncovers their indispensable yet often overlooked involvement in managing instruments at the frontier of their capabilities.

Keywords Mechanics · Materiality

In 1857, the first purchase order made by the newly-founded National Astronomical Observatory from the German manufacturer Repsold and Söhne, a universal instrument outfitted with micrometrical microscopes, arrived in an unusable state. Director Carl Moesta complained at the time that "because of the particular care with which this instrument has been built it would have been extremely useful for the observatory" but unfortunately, as he put it, "due to an uncommon and difficult to explain mishap" several of its pieces broke during shipping. Moesta's conclusion was clear: "As no artist is present in the capital from whom we can commission these repairs, the aforementioned instrument has been sent to the selfsame artists in Hamburg for this end."[1]

The use of the word "artist" here is interesting, as it surely refers to the German *handwerker*, which can also be translated as "artisan." Etymologically, "artisan" is someone who practices a manual art, as in German, but the German word emphasizes working with one's hands. Here the meaning of the term denotes a type of production that transforms and processes materials, but also refers back to the history of its adoption during the Industrial Revolution, which gave it a meaning associated with repair work.[2]

---

[1] Archivo Nacional de Chile, Memorias del Observatorio Astronómico Nacional, May 20, 1857.

[2] See: Der deutsche Wortschatz von 1600 bis heute. https://www.dwds.de. Accessed 5 March 2024.

© The Author(s), under exclusive license to Springer Nature Switzerland AG 2025   49
C. Sanhueza-Cerda, *The Day Laborers of Science. Technical Work at the Astronomical Observatory of Chile (1852–1927)*, SpringerBriefs in History of Science and Technology, https://doi.org/10.1007/978-3-031-84350-1_4

As laid out in the introduction, the work of technicians allows us to understand the role of materiality in the stabilization of knowledge. The work of dismantling instruments not only required workers who possessed "the skills needed to carry it out" but also objects that had certain properties making them "able to be dismantled."[3] Here, technical work takes on another meaning to the extent that instruments not only function through protocols, norms and directives, but above all through the object's own field of vulnerability, which required the intervention of these workers who have largely gone unstudied despite their essential role. In order to pull back the veil on this story, we will begin with their biographies so as to then explore actions and conflicts within the observatory regarding their responsibility for objects that were always at the limits of their possibilities.

4.1   Luis Grosch (-1902)

It's symptomatic of the visibility of technicians that we know so little about Luis Grosch despite the fact that he worked at the National Astronomical Observatory of Chile for over forty years. We know where he came from (Germany) and when he died, but not when he was born and, as this book goes to press, no images of him have been found. It seems that Grosch arrived as a settler to southern Chile in 1852, as one passenger manifest lists a Ludwig Grosch from Kassel, mechanic by trade.[4]

According to Friedrich Ristenpart, in the report he wrote for German astronomers in 1910, Grosch started working for the observatory in 1883, but the institution's own records show him as being active since 1853, at the very beginnings of the National Astronomical Observatory.[5] Here we can corroborate Shapin's argument that reviewing institutional documents for the payment of services is an ideal way of uncovering the work of technicians[6]: two years after the Chilean observatory was founded, we see Grosch being named the mechanic in charge of "the proper repair and conditioning of all of the observatory's instruments and of all the machines and physical and mathematical devices at the National Institute."[7]

It's interesting to observe the way in which technical expertise was evaluated by the University of Chile when trying to avoid paying for poorly performed or insufficient services. Technical work becomes visible precisely when it is evaluated negatively, which allows us to glimpse the work of technicians beyond the notions of "repair" or "reconditioning." When he was hired, it was established that Grosch

---

[3] Idem Der deutsche Wortschatz von 1600 bis heute. https://www.dwds.de. Accessed 5 March 2024.

[4] Held (1970, 42–46).

[5] Friedrich Ristenpart, "Los astrónomos alemanes en Chile," in: *Los alemanes en Chile*, Imprenta Universitaria, 1910, p. 11.

[6] Shapin (1989).

[7] Archivo Nacional de Chile, Memorias del Observatorio Astronómico Nacional, N°186 (74), January 7, 1857. Here we can see how there were attempts to incorporate technical work into an educational space such as the National Institute from the very beginning.

would be "required to carry out, without any other remuneration, the repair of the (...) mechanical instruments and devices when this does not amount to more than twenty pesos; if the value exceeds this quantity, he will be paid a conventional price." If there was no clarity regarding the cost of his work, a commission "composed of the observatory director, a university delegate and a competent person chosen by the interested party will decide, in the event of any doubt, if the value of the repair exceeds twenty pesos."[8] When Moesta took over as director, he requested that Grosch prepare a report on his activities as he had observed, in his words, "a great carelessness and lack of good faith in the performance of his duties." Grosch, for his part, demanded extra payment for the work he had performed, defending his repair of a heliostat in early 1854. For Moesta, this did not constitute a repair job: "For this adjustment, the mechanic did nothing but change the direction of movement of the mirror, adapting it to our hemisphere: something that surely was not worth twenty pesos, and for which he could not demand remuneration according to the clauses of the aforementioned decree." As we shall see, disputes between astronomers and mechanics formed part of the observatory's dynamics. Who decides if an instrument had been subjected to a repair? Is changing the direction of a mirror not a technical task? Would this instrument that was designed for the parameters of the northern hemisphere have been able to function without making such an adjustment?

The decision to repair or "fix" a device did not always favor the German technician living in Chile. As mentioned at the beginning of this chapter, Moesta preferred to return a device that had been sent from Germany as he believed that the manufacturer had handled it improperly. Moesta decided to pack "the aforementioned instrument and its broken pieces in the same boxes in which it had arrived and send it to Valparaíso, from where it has since embarked in recent days." In Moesta's report to the Minister of Public Instruction, his reasons were clear: "The mishap suffered by the instrument seems to be due (...) to the way in which it had been packaged; in such circumstances, nobody but Repsold could determine the cause of the breakage of the aforementioned instrument."[9] Here, Grosch played no role. This shows that expertise had its limitations and there wasn't always a place for it among the many precision and optical instruments of an astronomical observatory: Where was Grosch, then, what was his place? What did his work consist of? Was the repair or adjustment of instruments seen as a job?

With Grosch, we can see the extent to which the assembly of new devices out of the parts of others that had fallen into disuse, as well as the fabrication of entirely new ones, formed part of technical work. Once again, the discussions regarding payment for work performed provide clues to Grosch's place as a technician. In Moesta's aforementioned 1857 report, there is a discussion of the value of a "new electrical device that Mr. Luis Grosch has built for the institute in 1854 and for which he demands 18 pesos." Moesta recognized that while "the aforementioned clause of

---

[8] Archivo Nacional de Chile, Memorias del Observatorio Astronómico Nacional, N°186 (74), January 7, 1857.

[9] Archivo Nacional de Chile, Memorias del Observatorio Astronómico Nacional, vol. 84. June 15, 1857.

the decree does not mention the construction of new instruments but instead repairs (...) on several occasions we have made arrangements with this gentlemen so that he might make some simple device or another that we lacked at the institute's physics department." He assured the technician that he would be paid "the costs incurred for the metalworking or carpentry" but wouldn't pay him anything "for mechanical work," as was the case with an "instrument for measuring the direction of the wind." Regarding the payment for the electric lamp mentioned above, it was determined that the astronomical observatory would only assume the cost of "a table with two small mahogany columns, whose manufacture would cost eight pesos at most."[10] Ignacio Domeyko, Chair of the Faculty of Physical and Mathematical Sciences at the University of Chile, informed Moesta that "the lamp in question was made by Luis Grosch in March 1855 and [has been in use] since that time to illuminate the extended wires in the field of vision of the meridian circle."[11] Nevertheless, Domeyko agreed that the Chilean government should only cover the costs incurred and not the fabrication of the device itself. From the payment records, we know that Grosch created an instrument for measuring the wind shortly after he was hired and a standard barometer in 1875, both used by the observatory's meteorology department. In 1860, Grosch sold the observatory a reflection circle of his own manufacture, which was used to measure the angles between celestial bodies (up to 180°) and which Moesta used for the astronomy course he gave at the University of Chile.[12]

What were the limits of this innovation process? Did the use of an instrument for research purposes give it a different status? Adjust, repair: did these tasks not deserve to be recognized as creative acts? Or more concretely, the extra payment demanded by Grosch on more than one occasion?

Over the course of his career as a technician for the National Observatory of Chile, Luis Grosch manufactured precision lenses for the Breithaupt theodolite in 1859, added a position micrometer to the Dolland telescope in 1865, remade a clockwork piece for the Schadewell regulatory barometer that had arrived broken from Dresden in 1873 and redefined the Eichens comet seeker that had come from Paris in 1882 by building an appendix to ensure that the axis of the hour circle had the proper inclination, as it hadn't been built for Chilean latitudes, along with adding another piece to secure the axis of the hour circle in its rotations. It has been argued that certain maintenance practices are not oriented toward the "transformation and modification of objects" but rather their restoration: that is, "the continuation of their immutable being."[13] Maintenance work, in Grosch's case, allowed for the continuity of the observations of the heavens, or more precisely, the stabilization of materials that were constantly deteriorating and degrading. Grosch's work, in this sense, involved confronting the vulnerability of instruments that, like clouds on the night of a long-awaited astronomical event, forever threatened to undo everything: that moment

---

[10] Archivo Nacional de Chile, Memorias del Observatorio Astronómico Nacional, May 20, 1857.

[11] Archivo Nacional de Chile, Memorias del Observatorio Astronómico Nacional, May 20, 1857.

[12] Archivo Nacional de Chile, Memorias del Observatorio Astronómico Nacional, vol. 84, October 11, 1860.

[13] Denis and Pontille (2015, 360).

when observation nights, tedious measurements and enormous investments all led to nothing due to equipment malfunctions.

We know that Grosch participated as an observer during the April 25, 1865 solar eclipse and even sent a report on this astronomical event to Karl Moesta for publication in the German journal *Astronomische Nachrichten*.[14] Grosch's career in Chile was distinguished by his work establishing coordinates on geographical expeditions, as well as his surveying work during the construction of the Santiago sewer system. Despite the scarce biographical data we have on him and even the lack of images, his memory has become part of history: an anonymous motor of early Chilean astronomy.

4.2 Richard Wüst (1880–1954)

The observatory lacked technical staff following Grosch's death in 1902 and so its instruments deteriorated, making it very difficult to perform observations. When the director at the time, Friedrich Ristenpart, wrote to the German manufacturer Repsold regarding the construction of the meridian circle he had ordered for the Chilean observatory, he also inquired about the possibility of hiring a German technician, "preferably single," who would be offered a salary in accordance with the cost of living in this country where "it is not expensive," as well as lodging at the observatory. Ristenpart's conditions also required that the technician be skillful (*tüchtig*) but above all that he be a man of impeccable character (*tadellosem Charakter*), someone he could trust.[15] Here it becomes clear the extent to which maintenance and repair work were the precondition for observation and that it was therefore essential to have skilled and trustworthy technical staff.

Unbeknownst to Ristenpart, while he was negotiating with the Hamburg manufacturer, the Chilean government had hired another German who would become a cornerstone of the astronomical observatory: Richard Wüst. This technician, who had experience working for the German optics manufacturer Zeiss, had been recruited in 1909 as part of President Pedro Montt's policy of promoting Chilean astronomy. As with Grosch, we have information on his contract and occasional mentions of his work in the observatory director's annual reports, but we also have family photos and documents that have been kindly donated by his descendants, as well as the aforementioned correspondence between Ristenpart and Repsold, in which his work was repeatedly mentioned. We can thus sketch a more complex picture of Wüst and explore the everyday activities that tend to go unmentioned in astronomy and its histories. As Wüst was involved in the observatory's move from the Quinta Normal to Lo Espejo, we also have a record of his work because of the many difficulties this implied for the proper functioning of its instruments.

---

[14] Moesta (1865, 300).

[15] Letter from Ristenpart to Repsold, February 21, 1909, Hamburg State Archive, A II 28. Part of this discussion has been published in Isis. See: Carlos Sanhueza-Cerda (2022).

While we lack his original 1909 contract, three years later Ristenpart argued that it be renewed given the convenience of "retaining Mr. Wüst in the country for as long as possible, and even for him to remain here forever." As a precision mechanic, he was even essential to work beyond the walls of the astronomical institution. Ristenpart wrote to the Minister of Public Instruction that Wüst "has not only carried out many tasks for the observatory that are not commonly assigned to a precision mechanic, but the Seismological Institute also uses his services in difficult cases, as does the Army General Staff in order to repair and modify their complicated topographic instruments." According to Ristenpart, the creation of an astronomical observatory at Cerro San Cristóbal, a project by the United States that had been underway for nine years, "has taken advantage of his skill more than once and on one occasion he repaired the clockwork mechanism of the great refractor, which no other mechanic had been able to fix."[16]

One aspect emphasized by Ristenpart involved the savings to the Chilean state by not having to bring in foreign engineers to install the telescopes and measurement instruments that the country acquired. Precision had reached a point at which technical knowledge had to be imported alongside the devices themselves. In his report to the minister, Ristenpart mentioned that "to install these great telescopes in their wings of observatories, manufacturing houses tend to send one of their engineers, whose round trip travel expenses must be paid alongside their per diems." Here, "Mr. Wüst has left the Heyde astrophotographic equatorial (...) perfectly installed, saving the Treasury an important sum, as the offer of the Heyde firm in Dresden to send an engineer for this end was not needed."[17]

Wüst frequently had to correct the work done by local technicians who lacked experience in optics, which could seriously affect the stabilization of the instruments. Such was the case with the assembly of the dome for the aforementioned Heyde refractor, which had been assigned by the General Department of Public Works to a cement contractor. Ristenpart reported that it had been done "so deficiently that it later became necessary to carry out several modifications that were likewise unsuccessful, and so Mr. Wüst will be obligated to correct the defects of the dome."[18]

Wüst's arrival in Chile was good news to Ristenpart, which he shared with the Hamburg manufacturer, as the mechanic had not only managed to adapt pieces for observation in the Southern Hemisphere (a problem that had yet to be resolved in the early twentieth century) but had also managed to quickly dismantle the old Repsold refractor, despite not yet having a workshop of his own. This instrument, like many others, had gone unused for many years because nobody knew how to repair it. Ristenpart enthusiastically mentioned the possibility of building a workshop for Wüst as the country's precision instrument needs could require him in other

---

[16] Archivo Nacional de Chile, Memorias del Observatorio Astronómico Nacional, vol. 3105, December 5, 1912. Regarding the San Cristóbal Observatory, see: Silva (2019).

[17] Archivo Nacional de Chile, Memorias del Observatorio Astronómico Nacional, vol. 3105, December 5, 1912.

[18] Archivo Nacional de Chile, Memorias del Observatorio Astronómico Nacional, vol. 3105, December 5, 1912.

government departments, such as the navy, or even the private sector.[19] In a letter written the year after Wüst's arrival, Ristenpart discussed how the advice of the German technician had made him desist from purchasing replacement parts for the micrometer of the Gautier refractor's disk recorder; Wüst had argued that they were "expendable or rather superfluous."[20] Clearly, the guidance of technicians made it easier to understand what was required, what had to be acquired and what had to be modified so that the instruments functioned for the measurement tasks the observatory had committed to undertake. The possibility of having technical support also meant being able to better coordinate the observatory's needs with the manufacturer, not to mention purchase orders or requests for replacement parts for other instruments besides the meridian circle that was then being built.[21]

Having Wüst on hand allowed the observatory's instruments to be put to new uses. In his 1912 report to the Minister of Public Instruction, Ristenpart mentioned that "it will be the job of our mechanic (…) to repair and modify the old instruments that have been substituted by new ones and that currently function quite deficiently. This type of work can only be performed by a very competent person with a profound understanding of the task at hand, such as Mr. Wüst, and once completed, these instruments will be able to perfectly serve the observatory."[22] This task of reconditioning telescopes and measurement and photographic devices was key to the observatory's functioning.

In 1911, Wüst improved the installation of the contact in the Kessel sidereal clock that made it function, allowing it to be used for the first time in years.[23] The Gautier astrophotographic equatorial that had arrived from France in 1887 in order to be used for the international Carte du Ciel project had never been fully put into operation, which meant that Chile failed to take on the observations assigned to it. In 1895, Observatory Director Hubert Obrecht noted that the first astronomer had set out to perform a series of tests and observations in order to determine the equatorial's constants and verify its exact placement, as they had been unable to fully use it. The construction of a new observation site at Lo Espejo at the beginning of the twentieth century gave the instrument new life. Under Wüst's supervision, the dome of the astrophotographic wing had been readied for use in April 1911 and the first test observations were made soon thereafter. Wüst removed the eyepiece for complementary tasks and the telescope was definitively installed in July 11 of that year, producing its first useful plate on August 11. This plate was only obtained after Wüst had repaired the clock's irregular functioning through the use of a wing regulator with a special brake, later replacing it with a new one manufactured by

---

[19] Letter from Zurhellen to Repsold, August 21, 1909, Hamburg State Archive, A II 28.

[20] Letter from Zurhellen to Repsold, January 26, 1910, Hamburg State Archive, A II 28.

[21] Ristenpart had Wüst's help in tackling the problem of the Gautier refractor: not only did it need a new micrometer (which would modernize the device) but its clock also needed to be adjusted. Here the role played by Wüst becomes clear, as he was able to identify the problem with the device, allowing for the control of its second hand to be improved, correcting its tendency to lag. See: Letter from Ristenpart to Repsold, March 16, 1910, Hamburg State Archive, A II 28.

[22] Letter from Ristenpart to Repsold, March 16, 1910, Hamburg State Archive, A II 28..

[23] Archivo Nacional de Chile, Memorias del Observatorio Astronómico Nacional, vol. 3184, p. 8.

Gautier. In his 1911 report to the Minister of Public Instruction, Ristenpart stated that Wüst had built a device that allowed for photographic plates to be easily examined so as to not use the "primary measuring device" for secondary tasks and built two additional measuring devices "so that the measurement of plates can be done at a later time using three devices together." The Molyneux clock, another part of the instrument previously used to give the mean solar time, was dismantled, and cleaned by Wüst to be installed "in the astrophotography tower so that it could serve as a sidereal clock." Ristenpart concluded by stressing that Wüst had turned the Gautier refractor into a modern instrument by "correcting the errors of its initial installation at the Quinta Normal."[24]

The assembly and installation of the instruments that arrived in Chile was another of Wüst's responsibilities at the observatory. This not only involved working with optical and mechanical pieces, but also building domes that could move in conjunction with the movement of the instruments. A Bamberg transit telescope imported from the Friedenau-Berlin factory in Germany in 1911 had arrived in generally good condition, but some of its cords had been cut during shipping. Wüst created a new network for the instrument and installed it in the three-meter-square booth he built for it over the course of nine months. According to Ristenpart, this instrument was essential for determining the time at the new complex being built at Lo Espejo.[25] Wüst supervised and carried out the necessary tests for the installation of the Heyde equatorial telescope that had arrived from Dresden in 1912. This instrument required construction work on its wing of the institution, such as the installation of a dome, 8.1 m in diameter, that had also been sent by its German manufacturer.[26]

Many auxiliary pieces that were nevertheless necessary for the proper functioning of telescopes and measurement devices were made by Wüst. Such was the case with the device that allowed for up to three grids to be placed in front of a telescope's sight, which was to be used in the meridian department. One fundamental aspect for certain telescopes, such as the meridian circle, was to have a point of reference for calibration. In 1911, Wüst built a lithographic device at Lo Espejo to trace star charts on lithographic stones with mathematical exactness.[27]

In the aforementioned 1911 report to the minister, Ristenpart not only described the technician's work, but also discussed the mechanical workshop that Wüst had built for himself. This workshop contained pieces from the Berlin firm of Beling and Lübke and was equipped with a parallel lathe with a guide spindle and two interchangeable spindles, which had been previously tested at the *Physikalisch-Technische Reichsanstalt* in Berlin, as well as an instrument for marking mathematically precise

---

[24] Archivo Nacional de Chile, Memorias del Observatorio Astronómico Nacional, vol. 3184, pp. 13–15.

[25] Archivo Nacional de Chile, Memorias del Observatorio Astronómico Nacional, vol. 3184, p. 13.

[26] Archivo Nacional de Chile, Memorias del Observatorio Astronómico Nacional, vol. 3105, p. 38.

[27] Archivo Nacional de Chile, Memorias del Observatorio Astronómico Nacional, vol. 3105, p. 38.

divisions in metal, sent by the *Société Genevoise de Construction d'Instruments de Physique.*[28]

Did such support and dedication to the observatory's proper functioning mean that early-twentieth-century technicians were respected in Chile? Had the work of these secondary actors achieved greater recognition, given its evident importance not only for the observatory, but for any government institution that required precision work? Wüst stayed on after Ristenpart was gone. In the 1925 renewal of his contract, he is named as a "precision mechanic" at the Chilean observatory and is further assigned the work of "repairs at the Physics Department of the Pedagogical Institute."[29]

As we have seen with Grosch and Wüst, the scope and importance of the work of technicians was never remunerated in accordance with its impact. In the renovation of Wüst's contract in 1912, Ristenpart even requested that he be granted "the title of *mechanical engineer*, given the important work that he has performed and will perform later on, and because he will be obligated to carry out the same work as that performed by engineers in the great manufacturing houses of Europe."[30] We know that this title was never granted and that Wüst had to turn to side work to secure additional income. The invisibility of technical work clearly put a ceiling on the remuneration of manual labor. In early-twentieth-century Chile, as Zilsel has argued for other historical epochs, the tongue was above the hand and the word "above the thing."[31]

References

Boletín de Instrucción Pública. 1925. *Anales Universidad de Chile*, 602–603. Sesiones Agosto.

Denis, J., and D. Pontille. 2015. Material ordering and the care of things. *Science, Technology, & Human Values* 40 (3): 338–367.

Held, E. 1970. *Documentos sobre la colonización del Sur de Chile: Bosquejo histórico, nómina de barcos y personas que llegaron entre los años 1840–1875: De la colección histórica de Emilio Held Winkler*. Santiago: Claus von Plate.

Moesta, K. 1865. Schreiben des Herrn Prof. Moesta, Dir. Der Sternwarte in Santiago de Chile an der Herausgeber. *Astronomische Nachrichten* 1555, 300.

Sanhueza-Cerda, C. 2022. Stabilizing local knowledge: The installation of a Meridian circle at the National astronomical observatory of Chile (1908–1913). *Isis* 113 (4): 710–727.

Shapin, S. 1989. The invisible technician. *American Scientist* 77: 554–563.

Silva, B. 2019. *Estrellas desde el San Cristóbal*. Santiago: Catalonia.

Zilsel. 2008. *El genio. Génesis de un concepto*, Madrid: AEN.

---

[28] Archivo Nacional de Chile, Memorias del Observatorio Astronómico Nacional, vol. 3184, pp. 13–15.

[29] Anales Universidad de Chile (1925, 602–603).

[30] Archivo Nacional de Chile, Memorias del Observatorio Astronómico Nacional, vol. 3105, December 5, 1912.

[31] Zilsel (2008, 14).

Chapter 5
Mechanics and Astronomers in Competition: Who Validates Stabilization?

Abstract This chapter explores the conflicts and disagreements that arose between astronomers and mechanics spanning from the mid-nineteenth to the early twentieth century. Through an examination of scientific and public controversies, it investigates the interplay between the upkeep of telescopes and precision instruments and the advancement of astronomical understanding. This text examines the significance of recognizing the contributions of technicians in scientific endeavors, particularly in astronomy, where disputes arise over instrument functionality. Instances such as the controversy surrounding the National Astronomical Observatory of Chile highlight tensions between astronomers and technicians, revealing the intricate dynamics of scientific practice and the crucial role of technical maintenance.

Keywords Maintenance · Mechanics · Scientific controversies

Giving voice to *the hands* means seeking out those places where they've left behind traces of their work, such as disputes and polemics regarding the functioning of instruments. Routine work involves subjecting objects to rigorous tests of their maintenance and conservation that often lead to tensions between astronomers and technicians. As with the dispute between "amateur" and institutional astronomers over the imperfections of large telescopes in Great Britain at the end of the nineteenth century, the dispute over the state of the National Astronomical Observatory of Chile's instruments speaks to the dynamics of scientific practice.[1] It will thus be the lens through which we can observe the limits of technical work at the Chilean observatory.

Defective instruments derailed scientific work and needed to be repaired so that it could resume its normal course. As mentioned in the introduction to this book, it is precisely during moments of imperfection and maladjustment when technicians and mechanics in charge of maintenance and repair become visible and play a leading role. In this sense, both disputes as well as the maintenance and repair work itself provide a window for understanding and assessing the work of technicians. We will see how Grosch and Wüst, whose biographies we have already reviewed, confronted

---

[1] On the disputes over telescopes in Great Britain, see: Lankford (1981).

© The Author(s), under exclusive license to Springer Nature Switzerland AG 2025   59
C. Sanhueza-Cerda, *The Day Laborers of Science. Technical Work at the Astronomical Observatory of Chile (1852–1927)*, SpringerBriefs in History of Science and Technology, https://doi.org/10.1007/978-3-031-84350-1_5

the observatory's astronomers in the disputes that permeated public space at the end of the nineteenth century and the beginning of the twentieth.[2]

5.1   Despite What the Astronomers May Say, All the Instruments Are in Good Condition[3]

In 1885, several deputies from the National Congress requested that the Minister of Justice, Worship and Public Instruction provide information on the scientific work being performed at the National Astronomical Observatory and the productivity bonuses that its workers had received from the government as part of its responsibility for monitoring the use of public funds. No response was made to the request that year, nor was the issue discussed in Congress. On February 3, 1886, Conservative Deputy Manuel Balbontín (1845–1918)[4] insisted that the Minister of Public Instruction, who had final institutional responsibility for the observatory, give an accounting of the state of the observatory "as well has how much has been spent on instruments and parts over the past year."[5] In his view, the scientific institution was in a deplorable state that had come to overshadow the international leadership it had possessed under previous administrations. It can be argued that this complaint had less to do with a genuine interest in the nation's scientific activities than with political/ partisan interests: 1886 was an election year in Chile and the social context was permeated by constant suspicions of electoral fraud. The government was led by the lawyer Domingo Santa María (1824–1889), a member of the Liberal Party,[6] while the astronomical observatory was likewise led by a member of the Liberal Party, the geological engineer José Ignacio Vergara (1837–1889),[7] who had also been serving

---

[2] On the problem of technicians and technological trajectories, see: Dosi (1982), Cohen and Levinthal (1990) and Viotti (2002).

[3] Part of this work was published in Sanhueza-Cerda et al. (2020).

[4] Manuel Gregorio Balbontín Viancos was a lawyer who graduated from the University of Chile's Faculty of Law in 1870; afterward, he would exercise this profession as well as working as a journalist. In 1884, he was named General Director of the Conservative Party at its II National Convention. In 1885, he was elected the federal deputy representing Castro. See: Figueroa (1928, 70) and Bolados (1918, 95–101).

[5] Extraordinary Sessions of the Chamber of Deputies in 1885–1886, February 3, 1886 session, Santiago, Imprenta Nacional, p. 627.

[6] See: Portales (2004). Domingo Santa María was a lawyer who graduated from the University of Chile in 1847. He worked as a professor of geography, arithmetic and history at the National Institute in 1845, as well as at the University of Chile's Faculty of Philosophy. He served as a deputy, senator, mayor and government minister in different periods before being elected president in 1881, representing the Liberal Party. See: Figueroa (1901, 224–229, 1931, 780–782).

[7] José Ignacio Vergara Urzúa began his studies at the University of Chile's Faculty of Engineering in 1856 and received his degree in geographical engineering in 1863. He had been working as an assistant at the National Astronomical Observatory since 1860 and became its interim director in 1865, officially becoming its director in 1874. He was the President of the Primary Education Society from 1870 to 1874 and he helped found the Italian School. He was also a professor of

as the Minister of the Interior since 1885 and, in that capacity, was in charge of overseeing the electoral process.[8] And so investigating the government financing of the observatory was the perfect pretext to get at the man who would oversee the elections and thus affect the chances of the ruling party to remain in power.

Despite whatever personal or political reasons could have given rise to this public debate on the state of the observatory's instruments, the tenor it acquired as the months went by and its consequences for the institution's astronomers allow us to use it to analyze many key aspects of the role and expertise of technicians. Who could properly judge the quality and utility of a scientific instrument: the person who used it to make observations or the person in charge of maintaining and repairing it? Further still, did the political sector (which approved the institution's budget) have the capacity to judge the general state of a scientific institution and the quality of its production?

Initially, the question for these deputies involved the "physical state of the instruments." The Ministry of Public Instruction argued that this was due to a sustained budgetary shortfall,[9] but the deputies felt that the problem went deeper, arguing that scientific work was incompatible with political work. Guillermo Puelma Tupper (1851–1895),[10] a militant Radical Party deputy from Santa María and fierce opponent of the government, accused José Ignacio Vergara of "abandoning" the scientific institution in order to dedicate his time to political work for the government.[11] Vergara rejected this accusation and clarified that his job as director was *ad honorem*, but agreed with his accusers that administrative work was incompatible with scientific

---

mechanics, astronomy and differential and integral calculus at the University of Chile and one of the founders, alongside Ignacio Domeyko, of the Central Meteorological Office in 1868. He was awarded a medal for his meteorological observations at the 1875 Geographical Congress in Paris. See: Figueroa (1901, 433–434, 1931, 1029).

[8] Vergara also held a number of government positions while he was the director the National Astronomical Observatory, as there were no regulations preventing this and he had not been questioned for doing so until that time.

[9] Extraordinary Sessions of the Chamber of Deputies in 1885–1886, February 3, 1886 session, Santiago, Imprenta Nacional, p. 627.

[10] Guillermo Puelma Tupper studied medicine at the University of Chile and assisted in fighting the smallpox epidemic in Santiago in 1872, a labor for which he was decorated. In 1873, he became a member of the Academy of Fine Arts and published several poems. In 1876, he travelled to Europe in order to continue his medical studies at the universities of Heidelberg and Magdeburg, taking refresher courses in physiology. In 1878, he was the director of the Abraham Lincoln School and, in 1879, he entered the country's medical service, cofounding the House of Invalids during the War of the Pacific, involving himself in the Medical Society of Santiago and being named an extraordinary professor of histology at the University of Chile. In 1881, he began to engage in political journalism, editing and writing for the newspaper *La Época* until 1884, becoming known as a defender of Comte's positivism. He also wrote for the newspaper *La Libertad Electoral*. A member of the Radical Party, he was a deputy representing Parral from 1882 to 1885 and representing San Carlos from 1885 to 1888. He opposed the Domingo de Santa María administration and was known for his strong opposition to the Worship budget, as well as his anticlerical positions. See: Figueroa (1931, 571, 1897, 514–515).

[11] Extraordinary Sessions of the Chamber of Deputies in 1885–1886, February 3, 1886 session, Santiago, Imprenta Nacional, p. 627.

work as such. This was the reason why he claimed to not personally involve himself in astronomical observations. According to the minister/director, he had hired other astronomers to carry out these tasks: the second astronomer Wilhelm Wickmann in 1884 and the new first astronomer Adolf Marcuse (1860–1930), who would arrive in Chile in February 1886.[12]

As the months went by and the discussion advanced in the Chamber of Deputies, criticism deepened regarding the way the observatory managed its budget and they began to blame its director for not having spent on instruments or replacement parts. The lack of properly functioning instruments was considered to be the primary reason why the institution's scientific production was not at the level of foreign observatories, including those operating under similar conditions, such as the one in Córdoba, Argentina.[13] By August 1886, the judgment of the legislators had become devastating. In the words of Deputy Puelma Tupper: "Rarely have we seen a country's hopes for founding a scientific institute, created under such auspicious conditions, be so thoroughly defrauded."[14]

The institution's second astronomer, Wilhelm Wickmann, responded in the press. As an observatory employee, Wickmann could not directly respond in Congress, so the press became the ideal space for countering their arguments. Wickmann raised the problem of the authority and capacity of nonscientific actors (such as politicians) to judge the institution's activities. In the astronomer's eyes, subjects who lacked astronomical knowledge could not and should not judge the institution's work. To Wickmann, it seemed "very strange (...) that astronomical work, especially as it is quite original, must be presented for judgment to a public that is unable to understand much of these specialties." Even those who had a great deal of experience in astronomical matters in Chile, such as sailors and engineers, "need not concern themselves" with these matters. The astronomer felt that the "strangest thing" was "the judgment made by Mr. Puelma during the session of August 12. (...) He who wishes to issue a judgment on the value of astronomical work must not only be an astronomer, but he must also perfectly understand the quality and construction of the instruments in question."[15]

Here several points speak to expertise and how it is perceived as the basis for evaluating the proper or improper functioning of a precision scientific instrument.

---

[12] Adolf Marcuse was a German astronomer. He studied at the University of Strasbourg and in Berlin, obtaining his doctorate in 1884. In 1882, he worked as an assistant on the German expedition to observe the transit of Venus in the United States. He came to work at the National Astronomical Observatory in 1886 as its first astronomer. His contract was terminated at the end of that year. Following his departure from Chile, he worked at the Berlin Observatory from 1888 to 1891.

[13] Extraordinary Sessions of the Chamber of Deputies in 1885–1886, July 24, 1886 session, Santiago, Imprenta Nacional, p. 337.

[14] Extraordinary Sessions of the Chamber of Deputies in 1885–1886, August 12, 1886 session, Santiago, Imprenta Nacional, pp. 637–638.

[15] Guillermo Wickmann, "Rectificación al señor Puelma Tupper," *El Ferrocarril*, August 31, 1886, p. 2. The newspaper *El Ferrocarril* was founded in 1855, and by the end of the 19th century, it was the newspaper with the largest circulation in Chile. Ideologically, it was a liberal newspaper where many intellectuals and politicians published reflections and analyses of Chilean society. It was published until 1910. See Santa Cruz (2010).

Was the public *that is unable to understand much of these specialties* also unable to even opine as to whether or not a telescope was working properly? To Wickmann, the answer was clear: only specialists could form part of this discussion. And still further: not only did they have to be astronomers, but also someone who understood the *construction of the instruments in question.* So, a technician? Could a lawyer or a man of letters evaluate the work of scientists?

Despite Wickmann's attempts to keep the discussion within the spheres of astronomical and technical expertise, the debate continued to escalate in Congress. As the guardian of public funds, the legislature became yet another actor in the conflict. Parallel to these investigations and journalistic polemics, Liberal Deputy Jacinto Chacón (1820–1893) conducted an inspection of government-financed scientific institutions, which naturally included the astronomical observatory.[16] Chacón prepared a comprehensive report on the Quinta Normal de Agricultura, a space dedicated to cultivating the sciences and Chile's technological development; at the end of the nineteenth century, it included the National History Museum, the Botanical Garden, the Agricultural Institute and the National Astronomical Observatory.[17] For the latter, he received the support of the institution's recently-arrived first astronomer, Adolf Marcuse. According to Chacón, the purpose of his report was to "express the importance (…) of all of these establishments so that, having been understood and appreciated in all their value, they may be safeguarded by public opinion, Congress and the government, receiving all the resources they require."[18] Chacón's text was promoted as an example of how the scientific production of these institutions was not perceived as an isolated aspect of society, but as an essentially public matter; thus the importance of their auditing by the political world. Nevertheless, the question persisted as to the relevance of a disciplinary background when evaluating a scientific institution from the outside, in terms of the primacy of specialized knowledge. How could the issue of expertise be settled in such a way that it didn't invalidate the report?

One anonymous publication discussing Chacón's text on the Quinta Normal and its associated scientific institutions, which seems to have been written by Chacón himself, put forward an argument on the relevance of a legislator's assessment of the conditions at scientific institutions. "At first glance, the doubt arises that a man of letters and a commentator on our codified laws may not be the most appropriate person to write on matters that seem to be outside the orbit of his knowledge and routine labor in parliament and at a law firm," it read, insisting that this perception of his limits would be "a grave error." The text defined Chacón as someone who belonged

---

[16] Jacinto Chacón Barrios received his law degree in 1843. Also a poet, professor and journalist, he wrote for the newspaper *El Semanario de Santiago* in 1842 and *El Crepúsculo* in 1844. From 1851 on, he was an editor at *El Mercurio de Valparaíso* and was the director of *Revista del Pacífico* for the first half of the 1860s. Despite being profoundly religious, he was a Liberal Party member, being a substitute deputy for San Felipe from 1885 to 1888. See: Figueroa (1928, 517–518, 1897, 320–322).

[17] See Chap. 2 in this book.

[18] Chacón (1886, 4).

to "the reduced number of those who enjoy the enviable privilege of being well-versed not only in those matters immediately related to their professional work, but also many other spheres of knowledge." It argued that while Chacón's early literary career was distinguished "by his elevated inspiration as an original, productive poet (...) when the weight of age assayed its criteria (...) he became one of the country's most distinguished legal consultants (...) and a man of science who understands the formulas of algebra and trigonometry and recent advances in astronomy and geology, as well as the charms of the physical/chemical sciences." The text concluded by affirming what Wickmann had refuted: "In the author of the book in question (...) we see, then, an intimate and fortunate marriage between art and science."[19]

One obstacle had been overcome: a man of letters who possessed basic scientific knowledge could audit an astronomical institution. In fact, Chacón based a large part of his report on the state of its optics and precision instruments. The idea was simple and understandable outside expert circles: it was essential to ensure that instruments would give "the observer exact data for his astronomical deductions."[20] Chacón began his report with the assumption that astronomy "requires perfect and properly-installed instruments" as it is not an "inductive science (...) but on the contrary an exact, observed science that requires data in order to form its theories and mathematical calculations."[21] In other words, it was necessary to ensure that the observatory was functioning from its instruments on up. By centering the discussion on the proper functioning of the instruments, Chacón (perhaps without realizing it) echoed the requirements put forward by the discipline itself at the end of the nineteenth century, in which the fundamental problem of studying the stars revolved around observation techniques. On this point, the calibration of instruments and homogenization of data handling processes were fundamental, as were the social strategies undergirding these practices, such as administration methods, workplace organization schemes, etc.[22] The repetition of observation techniques and procedures at observatories all over the world not only meant the formation and professionalization of a scientific community,

---

[19] Anonymous, "La Quinta Normal. Sus establecimientos agronómicos i científicos," *La Época*, July 17, 1886, p. 3.

[20] Chacón (1886, 99).

[21] Chacón (1886, 99). These considerations are in concordance with the requirements of the discipline itself at the end of the nineteenth century. One fundamental aspect of the study of the stars was the expansion of observation techniques. Here, the calibration of instruments and the homogenization of data handling processes Bourguet et al. (2002), as well as the social strategies underlying these practices, such as administration methods, workplace organization schemes, etc. Aubin et al. (2010) and Raposo (2014) were fundamental. From this perspective, the repetition of observation techniques and procedures at observatories around the world not only meant the formation and professionalization of a scientific community Lankford (1997), but also the organization and control of its underlying observation networks Lankford (1981). At the same time, this allowed a discipline that had been legitimized by these protocols and proceedings to be differentiated from amateur astronomy Ogilvie (2000). These global requirements were also applied to South American observatories, such as those in Argentina (Minniti and Paolantonio 2009; Rieznik (2010, 2011, 2013) and Brazil (Barboza 2010).

[22] Bourguet, et al. (2002), Aubin, et al. (2010) and Raposo (2014).

but also the organization and control of its underlying observation networks.[23] At the same time, this allowed a discipline that had been legitimized by these protocols and procedures to be differentiated from amateur astronomy. The Chilean observatory was not alone in facing these problems, as similar questions and questionings arose in other South American countries, such as Argentina and Brazil.[24]

As we have seen, Chacón's pamphlet centered the debate, both in the press and in politics, on the so-called "exactitude of its instruments." This involved determining whether the instruments had been administered in a way that provided satisfactory results and whether they had contributed to global astronomical knowledge. On August 12, 1886, Deputy Puelma Tupper, delving into the work of the astronomical institution itself, requested that the Minister of Public Instruction compile all the manuscripts containing the observations made at the observatory, as well as of the work done up until that point by First Astronomer Adolf Marcuse, in order to develop an informed opinion. As he wrote: "All that has been revealed by the observatory's work published to date is the neglect with which it has been conducted and the inexactitudes that afflict it."[25] In particular, he questioned the condition of five instruments at the observatory: two equatorial telescopes, two meridian circles and a comet seeker. This political interrogation of the condition of the instruments led to technical discussions among members of the institution. The observatory's employees presented reports and counter-reports to the Ministry of Public Instruction, presenting their vision of the condition of their instruments, the quality of their scientific production and their opinions as to who held final responsibility for both. Now the members of the observatory itself would take center stage.

The first to intervene was the astronomer Adolf Marcuse through a report sent to the congressional investigatory committee, in which he stated that he had been unable to perform his work in a regular, exact fashion. This situation had also created difficulties for his colleague and countryman Wickmann, who, according to Marcuse, "categorically (…) has been unable to get any work done at the observatory for nearly two years due to this same cause."[26] Marcuse's report blamed mismanagement by Director Vergara for the underuse of the instruments and neglect of maintenance and repair work, which was the responsibility of the mechanic, Luis Grosch. Vergara responded categorically to this accusation: the problem didn't lie with the instruments but with the low scientific productivity of Marcuse, the observatory's first astronomer. In this dispute, Wilhelm Wickmann and Luis Grosch took the side of the director. Marcuse ended up confronting the powerful Vergara and was fired in the wake of this investigation: the first astronomer who had come from Germany in the hopes of turning the observatory around didn't last more than nine months in Chile.

---

[23] Lankford (1997), Rothenberg (1981) and Lankford (1981).

[24] For Argentina, see: Minniti and Paolantonio (2009) and Rieznik (2010, 2011, 2013). For Brazil: Barboza (2010).

[25] Extraordinary Sessions of the Chamber of Deputies in 1885–1886, August 12, 1886 session, Santiago, Imprenta Nacional, p. 447. This information was received by the Chamber of Deputies on August 17. *El Ferrocarril*, Santiago, August 18, 1886, p. 1.

[26] Archivo Nacional de Chile, Fondo Ministerio de Educación, vol. 554, "Representación del señor Marcuse dirigida al Supremo Gobierno," section II, undated.

The parties in conflict did not agree in their diagnostic of the problem nor as to who was responsible: there were different versions regarding the condition of the instruments, their capacity and the level of their use. Nevertheless, this dispute in the halls of Congress had a prehistory within the observatory itself, having emerged through circumstantial reasons before taking on a life of its own. It can be traced back to one year before the congressional investigation, in a letter written by Wickmann in September 1885 to the director of the National Observatory of Brazil, Luiz Cruls. While Wickmann remained loyal to Director Vergara in the congressional investigation, his correspondence speaks to prior conflicts at the Chilean observatory regarding the repair of instruments which involved the director, the astronomers and the mechanic Grosch. In his missive, Wickmann complained that the second equatorial telescope was old and useless because its filar micrometer was inoperable and its circle bent, while its annular micrometer was also in poor condition. He concluded his grievance by noting to the Brazilian astronomer that the Great Equatorial Telescope had gone practically unused but had been seriously damaged by its exposure to the elements, complete with bird droppings and rusty parts.[27]

Three months later, days before his colleague and countryman Adolf Marcuse joined the Chilean observatory as its first astronomer, Wickmann wrote a letter to him describing the poor state of its instruments, complaining that the mechanic Luis Grosch (who had been working at the institution for 32 years) had damaged the inclination of a meridian circle while cleaning its pivots, causing it to lose its cylindrical form. As the instrument could not be used for precise observations, Wickmann continued, he had requested that Director Vergara return the instrument to its factory in Europe in order to be repaired. Vergara refused, arguing that "things can't be done with such exactitude, at other observatories they don't do things so exhaustively either." Wickmann told his colleague that he had confronted the director, threatening to refuse to carry out any more observations using the instrument, and so Vergara ordered Grosch to examine it.[28]

Wickmann's letters to Cruls and Marcuse were presented to the investigatory committee by the latter in order to prove that even Wickmann, who supported Vergara and Grosch, distrusted the mechanic. The 1885 conflict thus led to a controversy over the relative knowledge of mechanics and scientists. In the letter Marcuse had received from Wickmann, the latter argued against the option of repairing the meridian circle in Chile by discrediting the expertise of the mechanic in his evaluation of the instrument's condition, explaining that he was "indifferent to whatever that man said or found, as it's not the custom in other countries to ask a cobbler, a tailor or a mechanic his opinion of astronomical matters, which are as important to me as that of a doorman."[29] Grosch defended himself by arguing that he had made a series

---

[27] Archivo Nacional de Chile, Fondo Ministerio de Educación, vol. 554, document no. 8, Letter from Wickmann to Cruls, September 20, 1885.

[28] Archivo Nacional de Chile, Fondo Ministerio de Educación, vol. 554, document no. 9, Letter from Wickmann to Marcuse, December 25, 1885.

[29] Archivo Nacional de Chile, Fondo Ministerio de Educación, vol. 554, document no. 9, Letter from Wickmann to Marcuse, December 25, 1885.

of adjustments to the instrument. Wickmann textually cited the mechanic's words: "The instrument is in excellent condition, everything Wickmann says is ridiculous, the observations made with this instrument are perfectly exact…!"[30] In his defense, Marcuse delivered a declaration that had been made by Grosch the following year, in which the mechanic argued that "the meridian circle in the right wing, or the *old circle* as others called it, is in a good state of repair and is in position to be used, as currently occurs."[31]

How was this controversy resolved? In the end, the debate regarding the state of the observatory's instruments and the final responsibility for the institution's deficient scientific production was settled politically, by national deputies and government ministers. In September 1886, Minister of the Interior and Observatory Director José Ignacio Vergara requested that the Minister of Public Instruction terminate the contract of First Astronomer Adolfo Marcuse. His reasons involved the scant work done in the seven months he had been at the astronomical observatory, as well as its poor scientific value. The astronomer was fired by the institution that same year, bringing this affair to its conclusion. Nevertheless, Marcuse had raised an issue that has troubled scholars of technology such as David Edgerton: those who do repair work are often situated in a position that allows them to question the uses of an object, or even its designers.[32] Is it possible that Grosch had generated new knowledge regarding the practices involving this instrument that Marcuse and Wickmann had been unable to understand? And who could best evaluate an instrument's functioning: the person who repaired and maintained it or the person who used it to generate scientific data? In this dispute, lawyers and men of letters had clearly put their thumbs on the scales.

5.2 It's but a Mere Invention of the Mechanic that I Don't Know How to Work My Devices

On December 12, 1912, the mechanic Richard Wüst submitted a complaint against Observatory Director Friedrich Ristenpart to the Ministry of Public Instruction. From the text, it's difficult to fully understand the situation: simply that Wüst had formed part of a scientific commission "along with the director, to Brazil," that the latter possessed "a selenium device he was unable to operate" and that, to use it in observations, he had required "the support of Professor Laub of La Plata." He then reported that Director Ristenpart had demanded that he "sign a document confirming the receipt of the sum of one hundred pesos ($100)" given that he had loaned the Chilean

---

[30] Archivo Nacional de Chile, Fondo Ministerio de Educación, vol. 554, documento no. 9, Letter from Wickmann to Marcuse, December 25, 1885.

[31] "Documentos relativos a los trabajos de este establecimiento remitidos por el señor Ministro de Justicia a la Honorable Cámara de Diputados y mandados a publicar el 17 del corriente" (1886), Santiago, Imprenta Nacional, p. 18. Cursive in the original text.

[32] Edgerton (1990) and Henke (1999).

government "a certain sum of money and had $150 in interest at the bank." Wüst declared that "he wasn't responsible for that situation." He concluded his accusation by reporting that "Mr. Cortinez (…) [argued] that I should apologize to him, which is not proper inasmuch as I was not guilty of anything."[33] After he made this statement, administrative proceedings were initiated against the director in connection to the matters "mentioned in this complaint." What was going on? As in the case discussed above, this was a combination of interpersonal issues, doubts regarding the financial decisions of Director Ristenpart and debates as to the quality of observations and technical control of instruments. Proceedings had not been initiated solely because Ristenpart and Wüst conflicted, nor could it be reduced to a scientific evaluation of his work. The intervention of the state, represented here by Manuel Gandarillas, who was in charge of the proceedings, occurred in the context of the oversight of the use of public funds, which in this case also involved an investigation into the "improper and indecorous conduct" of Ristenpart that had affected the reputation of Chile and its scientific institutions. As we shall see, given that part of these events occurred in Brazil, this case represented a problem for Chile's image abroad. We can also add the interventions of the press, recalling their role in the Grosch case, which speaks to the way scientific disputes appear as the explosive convergence of all of these elements together.

Here we must distinguish administrative aspects from those involving observations and instruments. In terms of the former, other witnesses clarified the issue over which Wüst confronted his boss: Ristenpart "wanted him to sign a receipt for travel stipends to Lo Espejo which he hadn't received."[34] Given the distance of the new observatory at Lo Espejo, the institution subsidized the train tickets of its workers commuting from Santiago. Ristenpart denied these accusations to Carlos Cortinez, the observatory's administrative director. The latter, in turn, confirmed that the director had asked the mechanic to sign receipts for money he hadn't received. Up until this point, it seems to be a purely administrative issue, but if it only concerned the institution's financial affairs, why did Wüst involve the issue of the use of selenium in the astronomical expedition to Brazil? It seems that the mechanic sought to establish that Ristenpart's "weak morality" was not only associated with payments to the observatory's staff, but also with the honesty of his own work as an astronomer. Following the course of these proceedings, the conflict began to encompass issues that brought into question Director Ristenpart's scientific prestige. This explains the mixture of precedents and declarations provided by the observatory's staff, as well as Ristenpart's defense, which centered on the efficacy and morality of his scientific and administrative decisions.

The context of this investigation, which exploded in Congress and the press, was a certain perception that Ristenpart was unable to handle problems of "staff discipline and morality," not to mention his relationships with his colleagues and the honesty of his own scientific research.[35] Ristenpart's contract was made possible thanks to

---

[33] Archivo Nacional de Chile, Fondo Ministerio de Educación, vol. 3105.

[34] Archivo Nacional de Chile, Fondo Ministerio de Educación, vol. 3105.

[35] Keenan et al. (1985, 133).

the support of President Pedro Montt as part of an ambitious program to revitalize the observatory, but the sudden death of Montt, who was ultimately his political protector, abruptly damaged what had seemed to be the beginning of a great career and a resurgence of astronomical progress in Chile after long years of difficulties and unreliable government backing. It didn't help that Montt's successor, Ramón Barros Luco, made major cuts to the astronomy budget, including the financing of the new observatory that was being built at Lo Espejo. In Congress, Ristenpart thus ended up confronting the conservative coalition that supported Barros Luco, which led the investigation to audit the observatory. As in the Grosch case, legislators settled their political differences by using the observatory as an electoral battlefield. The effect this had on public opinion, combined with the undiplomatic personality of the German astronomer, led the government to cancel his contract before it expired. For Ristenpart, this story would have a tragic ending.

The initial declarations in these proceedings had already mentioned the issue of selenium. According to Wüst, before Ristenpart embarked on the astronomical expedition to the Brazilian town of Cristina, he had requested literature on techniques for viewing a solar eclipse. The mechanic argued that, as the firm of Ernst Ruhmer had given instructions to the observatory director "that I couldn't contradict" and as "I didn't receive orders from the director to further explore this issue—which is otherwise purely scientific" he had left "everything in his hands."[36]

The use of selenium cells was an innovative technique that promised to improve astronomical observations, but had not yet taken root as a proven method in the field.[37] This perhaps explains why Wüst later declared that he was sure that Ristenpart himself "did not fully dominate the technique, and could therefore drag me into a disaster, as will be seen." To the mechanic, "this fear was entirely justified." Given Ristenpart's inexperience with these techniques, Wüst continued, the astronomer realized "the indispensable necessity of requesting the support of Professor Laub, as can be attested by Laub himself, Professor W. Knoche and the mechanic W. Trollund." This wasn't a complaint that Ristenpart had sought the assistance of his colleagues, which is necessary in any scientific practice, but rather a more serious problem, that of publishing the results of his observations without the consent of those colleagues who had participated in them: "Before the director of the Córdoba observatory could confirm (...) the results obtained by Mr. Ristenpart, they had already been published."[38] Ristenpart, in other words, had not only ventured to undertake an astronomical expedition that utilized techniques he hadn't mastered, financed by the Chilean government, but had also published his observations without consulting with his expedition partners or confirming his results, not to mention claiming authorship of what was actually a collective project.

---

[36] Archivo Nacional de Chile, Memorias del Observatorio Astronómico Nacional, vol. 3105, p. 6.

[37] It would be Richard Prager, who worked under Ristenpart as the observatory's second astronomer, who contributed to the astronomical use of selenium cells following his return to Germany and helped perfect photoelectrical measurement methods upon his arrival in the United States in the 1940s. See Hearnshaw (1993, 13–20).

[38] Archivo Nacional de Chile, Memorias del Observatorio Astronómico Nacional, vol. 3105, p. 6.

Ristenpart published the results of the expedition to Cristina, Brazil in an article in *Astronomische Nachrichten* whose title emphasized the new technique: "Observations During the October 9, 1912 Total Solar Eclipse Through a Selenium Cell." The text began by making it clear that the expedition had been financed by the Chilean government and had involved "the mechanic" (who went unnamed) and his assistant Rómulo Grandón (who, years later, would become the director of the Chilean observatory). While he mentioned that "the site (…) had also been chosen by astronomers in Córdoba under the direction of C. B. Perrine" and that he had collaborated with Dr. Knoche, director of the Meteorological Institute of Chile, "who had also come to Cristina in order to carry out aeroelectric measurements during the eclipse," everything seemed to indicate that it had been an individual effort. At most, he thanked Professor Laub of the University of La Plata in Argentina "for his advice and loan of a galvanometer," the latter of which was indispensable to the observations being conducted.[39]

At the beginning of his text, Ristenpart mentioned that he had ordered "four selenium cells, two flat and two cylindrical" from Ernst Ruhmer's firm in Berlin. He later mentioned that the flat cells had reached Santiago "in time to be tested there" while "the cylindrical cells only arrived during the trip." He then described the technique: "A cardboard tube is made, 5 cm in diameter and 1 m in length, in parallel to the long photographic telescope, and a flat selenium cell is placed at the inferior end of the tube." The cylindrical selenium cell "is installed vertically in the declination axis of the Steinheil telescope, its axis pointed at the sun." Reading this description, it would follow that the observations were made, yet later on he acknowledged that "all of these preparations were hampered, even ruined, by the inclement weather." Because of the rain, Ristenpart continued, "it became necessary to abstain from carrying out the established program." The problem thus lay not in technical expertise regarding the use of selenium, but with the weather. Nor was the abandonment of the original observation plan due to the improper handling of the instrument: Ristenpart mentioned that the device "was placed on the lower section of a workbench underneath a tree; the upper section largely protected it from the rain." Even though "a breakdown occurred some 17 min before the central unit, when the line was completely deenergized and the galvanometer's deflection dropped to zero," it finally "became possible to restore power when the galvanometer was substituted with one of the milliammeters and serial resistance was eliminated." According to Ristenpart, given that the galvanometer "started working again the next day following a leisurely examination, there was no internal defect in this instrument to which we can attribute its poor functioning."[40] In other words, observations had been made under acceptable and valid parameters and standards. His conclusions emphasized the innovative nature of these techniques: "Given that the number of prior observations made with selenium cells during total solar eclipses is not very large, it would be opportune to not underestimate the modest results of the Chilean expedition to Cristina, especially as they demonstrate the performance of selenium under such

[39] Ristenpart (1912, 418).
[40] Ristenpart (1912, 419).

unfavorable conditions."[41] Ristenpart considered what some could interpret as a failure as the oldest of scientific practices: trial and error.

On December 21, 1912, three months after the publication of his text in *Astronomische Nachrichten*, Ristenpart defended himself from the mechanic's accusations before Enrique Matta Vial, president of the congressional investigatory committee. In terms of his knowledge of the techniques used, while he acknowledged that "the use of selenium cells to determine changes in the intensity of light during an eclipse is rather new among astronomers," he made it clear that there were "two precedents," as well as "literature on their use in the observatory's library" which he had "duly (consulted) during my preparations for the eclipse," while also familiarizing himself with the "general qualities of selenium through a great physics text provided to me by Professor Ziegler."[42] Ristenpart said that it was in Ziegler's laboratory, in the presence of both the professor and the mechanic, where he had determined "the dark resistance of my flat cells" and conducted the practical tests "with the same milliammeters that were brought along on the expedition (…) sometimes in the presence of Mr. Prager, at other times alone." To Ristenpart, it was clear that it was all "a mere invention of the mechanic, to not use a stronger expression, *that I don't know how to work my devices.*"[43]

The director then sought to go beyond the accusation itself in order to reveal the intentions behind it, which he claimed only sought to disparage him and invalidate his scientific reputation. Ristenpart told the committee that he had brought along "four selenium cells, two flat and two cylindrical, while Mr. Knoche brought none and Mr. Laub brought one from La Plata, but of lower quality than mine." They then developed "a joint plan to make use of all five cells." What ended up happening, as Ristenpart had recounted in the article published in Germany, was that the atmospheric conditions had prevented them from conducting their observations the way they had planned: "Things ended up going differently. The plan had to be totally modified because of the rain. We could only utilize one cell with a direct observer, me." If it were not for this unforeseen circumstance, he continued, "the results would have been collected by the scientist with greatest seniority, he would have led the drafting of the report, which would then be passed along to the others in order to incorporate their changes and the publication would have been jointly signed by the three of us." In the end, there was "only one cell with a direct observer, and I worked under the rain with almost no protection from seven in the morning until six at night." After eleven hours of observation exposed to the elements, "while (according to Ristenpart) Knoche and Laub made their observations through an electrical device within the house," he drafted "the first summary of the observations made; the effects of the eclipse were so clear that I could telegraph the news of our success, achieved in

---

[41] Ristenpart (1912, 419).

[42] Archivo Nacional de Chile, Memorias del Observatorio Astronómico Nacional, vol. 3105, p. 92.

[43] Archivo Nacional de Chile, Memorias del Observatorio Astronómico Nacional, vol. 3105, p. 92. Emphasis in the original.

spite of the rain, not only to the Ministry of Instruction, but also to the astronomical journal *Astronomische Nachrichten.*"[44]

To the astronomer, the origins of this accusation could only have one explanation: "It is difficult for me to admit, but my enthusiasm only awoke the envy of my travel partner, Dr. Knoche." According to Ristenpart, Knoche "wrote to several of the German professors here upon his return to Santiago in order to advance his rights with regard to these observations, but they all rejected his motions" given that "it's a sacred law in science that each researcher has the right to publish his research under his own name, unless he was simply executing the orders of his superior." With regard to Laub, Ristenpart says that his colleague "addressed a letter from La Plata, along with a request that I send him a copy of my observations (…) I replied by asking in what fashion, according to his ideas, collaborative work was to be done between two scholars." Ristenpart's declarations to the investigatory committee make it clear that he felt "there was no collaboration, as Mr. Laub made no observations." While Laub had loaned equipment to the Chilean astronomer, the latter declared that "naturally he was grateful to Mr. Laub for his kindness and loan of the device. Nevertheless, it will be clear to all that these circumstances did not provide a basis to simply and plainly cede all of the results of my observations to Mr. Laub so that he can publish them under his own name."[45] Finally, Ristenpart declared that the results of his observations using selenium cells were publicly available ("with curves and calculations") "at the Historical Institute in Río, at the Scientific Society of Buenos Aires and the central salon of the University of Chile, and will soon be made available to the scientific world through *Anales de la Universidad*, *Anales de la Sociedad Científica Argentina* and, mostly importantly, *Astronomische Nachrichten*, a journal that is read by astronomers in every country. Exposed to the criticism of so many scholars, they can certainly likewise withstand the criticism of the mechanic, Mr. Ricardo Wüst."[46] The astronomer's attempts to "put him in his place," as with all those situated below him in the social hierarchy, were evidently not free from sarcasm.

Yet it remained unclear why Wüst had involved the issue of the payments for train tickets to Lo Espejo with the issue of the eclipse in Brazil, which, to Ristenpart, was "a fairly old affair." What had driven the mechanic to combine the two incidents? To Ristenpart, the answer went beyond scientific issues: Wüst had been carried away by "the words of Mr. Knoche" for political reasons. The director's conclusion was judgmental: "I am deeply sorry that this gentlemen has struck up a relationship with the mechanic, whose socialist ideas could not fail to escape him." In other words: there had been a transference of personal and even political conflicts within the observatory to the work of its astronomers.[47]

---

[44] Archivo Nacional de Chile, Memorias del Observatorio Astronómico Nacional, vol. 3105, p. 94.

[45] Archivo Nacional de Chile, Memorias del Observatorio Astronómico Nacional, vol. 3105, pp. 98–99.

[46] Archivo Nacional de Chile, Memorias del Observatorio Astronómico Nacional, vol. 3105, pp. 99–100.

[47] The director's suspicion that the mechanic participated in socialist ideas was part of the context of workers in Chile at the beginning of the 20th century. It has been said that workers were subjected to

It's interesting to observe how the press reflected these conflicts, acting as an echo that has come down to us as a historical record and testimony of the growing importance of astronomy in early-twentieth-century Chile. Public opinion had become so familiar with astronomers that the popular perception of Director Ristenpart's poor character was even captured in the weekly cartoon that the illustrated magazine *Zig-Zag* published on September 14, 1912: the enraged face of the German astronomer was included as part of an advertisement for the nerve tonic Gliceforosfato Robin (see Fig. 5.1).[48]

How was it possible for a scientist from a discipline so elevated above the masses to be so well known? The answer, in part, lies in the expectation that science could help predict natural phenomena that affected the population, such as earthquakes or tsunamis, or even astronomical phenomena, as with the 1910 approach of Halley's comet. As the public did not distinguish between disciplines, nor did it understand the geological origin of earthquakes, it was believed that the observation of the skies (which was the responsibility of the National Astronomical Observatory) was the key to understanding these phenomena. This explains something difficult to understand today, living as we are after the division of scientific disciplines: "To explain the possible effects of a certain celestial phenomenon or to calm the population in the wake of a prediction involving the stars, the expert voice was generally that of (…) astronomers."[49] This was something that Director Ristenpart understood very well. Given his prior experience in science popularization at the Berlin planetarium Urania, he organized public conferences, guided tours of the observatory's facilities and educational events for students and even boy scouts, as well as informing the press of their progress, discoveries and achievements. At times of public alarm, reporters would even wait for Ristenpart outside the observatory for information that would help them to understand the forces of nature.[50]

Ristenpart's relationship with the media, particularly the possibilities it provided to illustrated periodicals such as *Zig-Zag*, unquestionably speaks to his success as a popularizer of astronomers' activities behind the observatory's walls. Photographs of the construction of the new observatory at Lo Espejo were published alongside *Zig-Zag*'s reportage, revealing the trusses of the new buildings and the instruments themselves, along with a sort of collage of the director making observations and measurements, or even holding social gatherings with other astronomers working

---

intense politicization in the context of the emergence of leftist parties such as the Democratic Party of Chile and later the Communist Party of Chile. This influence is manifested in their incorporation into the ranks of the labor movement and the middle class, to which the mechanics who worked at the Chilean Observatory also belonged in one way or another. Regarding Chilean workers and socialist ideologies, see Grez (2011) and Grez (2016). For the Chilean middle class, see Candina (2013).

[48] On cartoons and astronomy in early-twentieth-century Chile, see Ramírez and Sanhueza-Cerda (2023). Zig-Zag magazine, founded in 1905, is considered the first miscellaneous publication produced in Latin America. It stood out for its variety of topics and its concern for design, graphic communication and visual quality. It was published until 1964. See Donoso (1950).

[49] Ramírez (2019, 240).

[50] See: Valderrama (2021).

Fig. 5.1 Revista Zig-Zag, September 14, 1912, Santiago de Chile

in the country at the recently-inaugurated US observatory at Cerro San Cristóbal (see Fig. 5.2).[51] Each action or improvement was communicated to the press: to promote a public conference or provide information on an exhibition, but also to show off the purchase of new instruments or advertise scientific expeditions and new publications. There was a story in the newspaper *El Mercurio* on the journey to Brazil

---

[51] On the US expedition, see: Silva (2019).

El nuevo observatorio astronómico.

El primer pabellón en construción.   Los asistentes á la colocación del primer tijeral
                                           del observatorio astronómico de «Lo Espejo»

Fig. 5.2   Revista Zig-Zag, August 12, 1911

that unleashed this conflict, published on December 5, 1912, which emphasized a talk on "the work of the Chilean expedition that observed the total solar eclipse in Brazil this past October 10." Here we learn that "the expected audience for this event occupied the ample venue." The article continued: "Mr. Ristenpart began by recounting the expeditions that had been sent by our government to observe identical phenomena and concluded by describing the preparations and results obtained by the expedition to Brazil." The article concluded by mentioning that the observatory director "illustrated his discussion with luminous projections of some interest."[52]

The director's interest in promoting his work and that of the observatory as a whole, while positioning him in Chilean society, also exposed him to public scrutiny. *El Mercurio* praised its astronomical work and the role of its director, but others, particularly *La Razón*, were critical of those issues that were left opaque by the institution's official channels. It was natural that the use of selenium on the Brazilian expedition would be taken up by journalists seeking to carve out a space for themselves in the public debate.

The spark that detonated this explosive situation was an article published in *La Razón* on December 20, 1912 with a disturbing title: "Grave Difficulties at the Astronomical Observatory." After describing the "malaise that has been observed at the Astronomical Observatory of Santiago for some time now" and the "tension between the director, Mr. Federico Ristenpart, and his staff," the article mentioned that "the observatory mechanic, Richard Wüst, often harried by the administration, has recently initiated an investigation into his superior, presenting a serious complaint

---

[52] El Mercurio, December 6, 1912, *El eclipse de sol último. La conferencia de anoche en la Universidad.* Ristenpart particularly emphasized images not only to promote the observatory, but also to show off the results of his own work. El *Mercurio de Santiago*, a conservative newspaper, is to this day one of the most relevant in Chile. It was founded in 1910, and in its beginnings, like many other newspapers of the time, it sought not only to influence politically but also to organize itself as a commercial and profit-making enterprise. See Ossandón and Santa Cruz (2001).

to the Ministry of Instruction in which he promises to reveal grave irregularities." The report, signed with the pseudonym Alfapeleo, reported that the mechanic did not wish to speak, but that another observatory functionary had anonymously reported that most of the accusations were true, and even went so far as to declare that the observatory "had done no work that deserved to be taken into account by our colleagues."[53]

Nine days later, a supposed interview was published in the same newspaper, though it was more of a journalistic ambush of Ristenpart. The article had allegedly been written "to allow the director to defend himself."[54] After a brief dialogue that repeatedly showed Ristenpart to be reluctant to respond to the charges against him and even mocked him for his poor Spanish and his German accent, the article proceeded to enumerate the accusations. Apart from the financial problems, he was criticized for his "absolute lack of skill and (...) the shameful failure of his observations of eclipses, etc.,"[55] alluding to the Brazilian expedition.

On January 1, 1913, *La Razón* once again questioned the work of the German astronomer, this time directly addressing the questions surrounding his scientific competence and his contributions to astronomical knowledge. The article made it clear that its intention was not "to affect him personally, but rather as the director of a center of culture and dissemination of our progress abroad. By enunciating Mr. Ristenpart's stumbles and errors, perhaps he shall abandon a position that requires qualities he disgracefully lacks." The issue of the quality of his observations reappeared in greater detail: "There was then the case of the recent expedition made to Brazil to observe the total solar eclipse in the state of Minas Gerais. Of the eclipse he saw nothing and the selenium readings that were taken after the phenomenon occurred were energetically rejected by M. Perey, the astronomer of the Córdova observatory." Clearly the problem of the quality of astronomical work would be judged by public opinion.

The issue of selenium resurfaced on January 5, this time involving his collaborators on the Brazilian excursion. The newspaper *La Razón* enlisted Walter Knoche, at that time the director of Chile's Central Meteorological and Geophysical Institute, whom the journalist interviewed "as to whether the celenium [sic] that Mr. Ristenpart had utilized to observe the recent solar eclipse in Brazil was defective or if his failure was due to its clumsy use." Knoche apologized to the journalist, claiming that he was "morally implicated for having served under the same banner as Mr. Matta Vial" and that any declaration he made "regarding this matter would prejudice the

---

[53] En el Observatorio Astronómico (primera parte) (La Razón, December 20, 1912) Article title: En el Observatorio Astronómico. Graves dificultades en este servicio. The newspaper La Razón was a publication associated with the Radical Party of Chile, of a progressive and social-democratic nature that echoed the discussions about the workers' and students' movement. See Santa Cruz (2003).

[54] The article textually read: "—Pero señog [sic], yo creo habeg [sic] contegtao [sic], mi ser amico [sic] de LA RAZON e pogque [sic] hacegme [sic] estas preguntas?".

[55] El Observatorio Astronómico. Graves cargos contra el sr. Ristenpart (La Razón, December 29, 1912) Article title: El Observatorio Astronómico. Graves cargos contra el Sr. Ristenpart.

proceedings currently underway and so the investigation would not deliver the true results it pursues."[56]

La Razón returned to the matter six days later, insisting on the question of Ristenpart's actions during the eclipse in Brazil and his use of selenium cells. The article began by citing a scientific report then being prepared by "the astronomers Laub of La Plata, Perrino of Córdoba and Knoche of Santiago," who "analyzed the tests that the director of our observatory conducted with selenium during the recent eclipse, concluding with a disavowal of the talks given in Buenos Aires" by Ristenpart on the aforementioned astronomical event. According to the article's sources, to ensure the success of the mission, "they should have carried out observations of humidity, air pressure, eclipse wind, etc., which was to be done by Dr. Laub's wife, but as it was raining copiously that day, Dr. Ristenpart gallantly offered to do it himself." Nevertheless, these observations "were done poorly, and Mr. Knoche, who had a special interest in verifying the results, has been unaware of them to this day, as has Mr. Laub."[57] The disavowal of the director's scientific abilities by his own colleagues was unanimous. In other words: Ristenpart didn't know how to do his job.

Three months later, *La Razón* reported on the "Tragic End of Mr. Federico Ristenpart": "When, in the performance of our terrible duty, we informed the government of the serious irregularities revealed in the administration of the astronomical observatory, we never believed that confirming these charges could lead to such an extreme situation." The newspaper sought an explanation for this tragic occurrence: "It seems that the energetic decision by the government profoundly altered this gentleman's nervous temperament, to the extreme at which his friends believed they had observed a mental disturbance during his final days." The consequences were described with an impersonal objectivity: "Yesterday morning, in his residence at the Quinta Normal, Mr. Ristenpart shot himself, dying instantly."[58] The case of the selenium was finally closed.

The dispute between Ristenpart and Wüst resembled the one between the mechanic Grosch and the astronomers working at the observatory at the end of the nineteenth century. In both cases, there's a conflict of expertise regarding the handling of instruments and diagnostics of their proper functioning. It's interesting to note an inversion of traditional hierarchies to the extent that technicians validated their arguments through their proximity to objects: while both sides were heard out, it was the word of the astronomers that was generally questioned. In a debate regarding objects, a condition that humbled technicians before scientists, as reflected in their pay and level of social prestige, nevertheless situated them in a privileged position. The field of workshops, replacement parts and calibrations continued to be the *locus* of mechanics, no matter how renowned astronomers were in their disciplines.

---

[56] Lo del Observatorio Astronómico (Segunda parte) (La Razón, January 5, 1913), Article title: Lo del Observatorio Astronómico. Siguiendo nuestras investigaciones.

[57] La cuestion de Lo Espejo (Segunda parte) (La Razón, January 11, 1913), Article title: La cuestión de Lo Espejo.

[58] El director del Observatorio Astronómico (2) (La Razón, April 10, 1913), "El Director del Observatorio Astronómico. Trágico fin de don Federico Ristenpart."

This explains the aggressivity of Wickmann and Ristenpart. Wickmann situated the mechanic at the level of a cobbler or tailor: how could the judgment of a scientist be questioned by the arguments of a manual laborer? The hand continued to be situated below the word. Meanwhile, Wüst's doubts regarding Ristenpart's handling of the selenium was nothing but *a mere invention of the mechanic*, one that was politically driven. Wüst, for his part, sought to dignify technical work through the value it had for scientific research. In his reports to the government minister, Wüst explained that Ristenpart did not give him sufficient time to learn to use selenium and thus be in a position to aid in the observation work in Brazil. Rather than basing his argument on Ristenpart's incompetence at techniques he hadn't mastered, Wüst argued that the expedition's failure was largely due to having been unable to make the proper technical preparations prior to the astronomical event. The idea of the *day laborers of science* becomes meaningful to the extent that Ristenpart's solitary efforts were a fiasco. Without the collaboration of technicians and astronomers, the skies were but shadows on a useless plate.

References

Aubin, D., C. Bigg, and H. Sibum, eds. 2010. *The heavens on earth: Observatories and astronomy in nineteenth-century science and culture.* Durham, London: Duke University Press.

Barboza, C. H. M. 2010. Ciência e natureza nas expedições astronômicas para o Brasil (1850–1920). *Boletim do Museu Paraense Emílio Goeldi. Ciências Humanas, Belém* 5 (2), 273–294.

Bolados, A. 1918. *Álbum del Congreso Nacional en su primer centenario 1818–1918.* Santiago: Imprenta España.

Bourguet, M. N., C. Licoppe, and H. O. Sibum. 2002. *Instruments, travel and science. Itineraries of precision from the seventeenth to the twentieth century.* London: Routledge.

Candina, A. 2013. *Clase media, Estado y Sacrificio.* Santiago: LOM.

Chacón, J. 1886. *La Quinta Normal y sus establecimientos agronómicos y científicos.* Paseo de estudio, Santiago: Imprenta Nacional.

Cohen, W., and D. Levinthal. 1990. Absorptive capacity: A new perspective on learning and innovation. *Administrative Science Quarterly* 35 (1): 128–152.

Donoso, R. 1950. *La sátira política en Chile.* Santiago: Editorial Universitaria.

Dosi, Giovanni. 1982. Technological paradigms and technological trajectories: A suggested interpretation of the determinants and directions of technical change. *Research Policy* 11 (3): 147–162.

Edgerton, D. 1990. *The shock of the old: Technology and global history since 1900.* London: Profile Books Ltd.

Figueroa, P. 1897. *Diccionario biográfico de Chile.* Tomo II. Santiago: Imprenta Barcelona.

Figueroa, P. 1901. *Diccionario biográfico de Chile.* Tomo III. Santiago: Imprenta Barcelona.

Figueroa, V. 1928. *Diccionario histórico, biográfico y bibliográfico de Chile.* Tomo II. Santiago: Balcells & Co.

Figueroa, V. 1931. *Diccionario histórico, biográfico y bibliográfico de Chile.* Tomos IV–V. Santiago: Imprenta La Ilustración.

Grez, S. 2011. *Historia del comunismo en Chile. La era de Recabarren (1912–1924).* Santiago: LOM.

Grez, S. 2016. *El Partido democrático de Chile. Auge y ocaso de una organización política popular (1887–1927).* Santiago: LOM.

Hearnshaw, J. B. 1993. *Photoelectric photometry. The first fifty years.* Cambridge: Cambridge University Press.

Henke, C. 1999. The mechanics of workplace order: Toward a sociology of repair. *Berkeley Journal of Sociology* 44: 55–81.

Keenan, P.C., S. Pinto, and H. Alvarez. 1985. *El Observatorio Astronómico Nacional de Chile (1852–1965)*, 1985. Santiago: Universidad de Chile.

Lankford, J. 1981. Amateurs versus professionals: The controversy over telescope size in late Victorian science. *Isis* 72: 11–28.

Lankford, J. 1997. *American astronomy: Community, careers, and power, 1859–1940.* Chicago: University of Chicago Press.

Minniti, E., and S. Paolantonio. 2009. *Córdoba estelar: Desde los sueños a la astrofísica: Historia del Observatorio Nacional Argentino.* Córdoba: Universidad Nacional de Córdoba.

Ogilvie, M.B. 2000. Obligatory amateurs: Annie Maunder (1868–1947) and British women astronomers at the dawn of professional astronomy. *The British Journal for the History of Science* 33 (1): 67–84.

Ossandón, C., and E. Santa Cruz. 2001. *Entre las alas y el plomo. La gestación de la prensa moderna en Chile.* Santiago: LOM.

Portales, F. 2004. *Los mitos de la democracia chilena. Tomo I. Desde la Conquista hasta 1925*, 1° ed. Santiago: Catalonia.

Ramírez, V. 2019. Expertos y profanos: Circulación del saber astronómico en magazines chilenos (1900–1920). *Revista de Humanidades* 40, 235–272.

Ramírez, V., and C. Sanhueza-Cerda. 2023. ¿Observadores u observados?: Los astrónomos bajo el escrutinio de ilustradores chilenos (1908–1919). *História Unisinos* 27 (3): 571–583.

Raposo, P. 2014. Time, weather and empires: The campos Rodrigues observatory in Lourenço Marques, Mozambique (1905–1930). *Annals of Science* 72 (3): 279–305.

Rieznik, M. 2010. El Bureau des longitudes y la fundación del observatorio de La Plata en la Argentina (1882–1890). *História, Ciências, Saúde—Manguinhos* 17 (3), 679–703.

Rieznik, M. 2011. *Los cielos del sur. Los observatorios de Córdoba y de La Plata, 1871–1920.* Rosario: Prohistoria.

Rieznik, M. 2013. The Cordoba observatory and the history of the personal equation' (1871–1886). *Journal for the History of Astronomy* 44 (156): 277–301.

Ristenpart, F. 1912. Beobachtungen wahrend der totalen Sonnenfinsternis des 9. Oktober 1912 mittels einer Selenzelle. *Astronomische Nachrichten* 194 (4656), 418.

Sanhueza Cerda, C. et al. 2020. Todos los instrumentos están en buen estado. Disputas en torno al funcionamiento de los telescopios del Observatorio Astronómico Nacional de Chile en el siglo XIX. *Asclepio. Revista de Historia de la Medicina y de la Ciencia* 72 (1), enero-junio, 1–11.

Santa Cruz, E. 2003. El campo periodístico en Chile a principios del siglo XX. *Periodismo y Sociedad* 14: 17–29.

Santa Cruz, E. 2010. *La prensa chilena en el siglo XIX. Patricios, letrados, burgueses y plebeyos.* Santiago: Editorial Universitaria.

Silva, B. 2019. *Estrellas desde el San Cristóbal.* Santiago: Catalonia.

Valderrama, L. B. 2021. *Todos los temblores después del terremoto. Configurar la experticia en un país sísmico.* Santiago: UAH Ediciones.

Viotti, E. 2002. National learning systems: A new approach on technical change in late industrializing economies and evidences from the cases of Brazil and South Korea. *Technological Forecasting and Social Change* 69 (7): 653–680.

Chapter 6
Collaborators

Abstract This chapter highlights the importance of collaborators at the National Astronomical Observatory of Chile, whose functions were distinct from those of mechanics and astronomers, but essential to the operation of the observatory. This case-based text sheds light on the hierarchical dynamics and contractual constraints within the institution, highlighting the fundamental role of labor practices in astronomy. It demonstrates how the contributions of collaborators, often overlooked, were crucial to the success of astronomical endeavors, challenging the notion of astronomers as the sole protagonists of scientific research. This chapter deserves a special section on the hidden role of women at the beginning of the twentieth century, when in Chile it was not possible or very difficult for women to develop academically and participate in the country's scientific activities.

Keywords Science collaborators · Calculators · Women in science

"In a country with a relatively young culture, such as Chile, the continued existence of the people as a whole is naturally the greatest concern. The individual mentality revolves around these concerns."[1] Richard Prager, the second astronomer at the National Astronomical Observatory of Chile, thus began a letter dated December 29, 1912 in which he tried to explain to Enrique Matta Vial the difficulties faced by Director Friedrich Ristenpart with his staff. This situation had even led to the crisis within the observatory that Matta himself was investigating as part of the administrative proceedings described in the previous chapter. How important could this conflict really be if, as we often tend to think, the workers at a scientific institution are practically *background characters*? It seems that Prager believed that the heart of this crisis (whose ins and outs we will explore below) revolved around this group, which reveals that the astronomers themselves understood that they were essential to the continuity and precision of their work. Given the social situation in Chile, which was marked by material questions of economic existence, Prager reflected

---

[1] Letter from Richard Prager to Enrique Matta Vial. Archivo Nacional de Chile, Fondo Ministerio de Instrucción Pública, vol. 3105, pp. 203–204.

that "the devotion sacrificed to scientific activity for its own sake is only found here on very rare occasions" and so "employees only work for their salary, not for the inner satisfaction provided by astronomical activity (…) they haven't come to the observatory due to an internal inclination but an external cause, and they would abandon astronomy tomorrow (…) if they were offered a better-paying job."[2]

A conflict that many had attributed to the personal characteristics of the observatory director was interpreted by his colleague Prager as being due to Ristenpart's vision. Astronomical activity produced "a great satisfaction" in the observatory director, Prager reflected, and so he thought that the same "should occur with all his employees." For Prager, "this also reveals his very peculiar tendency to only assume the best of people, without thinking to examine their real circumstances." And their *real circumstances* were precisely the conditions of wage labor in Chile. Following Prager's argument, what was going on here was no more and no less than a "conflict between the ideal and material conceptions of life." Transforming these labor conditions, then, would benefit the observatory. Prager even ventured to say that, in retrospect, "it probably would have been better to launch the observatory on a smaller scale, with fewer employees, and awaken and fortify the idealistic interest required by all scientific activity, but especially astronomy, over decades of work." Fewer employees with a greater interest in astronomy would be the key to the development and advancement of the observatory. Nevertheless, Prager believed that "a reduction in staff, while it could relieve some current difficulties in itself, could also lead to serious setbacks that I would profoundly lament unless accompanied by a simultaneous improvement in the qualitative performance of our employees."[3]

The perspective of this German astronomer regarding the country that had taken him in situates us on the everyday plane of the employees, collaborators and assistants who inhabited what the historiography tends to present to us as an exclusive place for astronomers. But who was considered to be a collaborator or a functionary? What differentiated a mechanic from a collaborator, or even from the astronomers themselves, who were often hired in this fashion? It would seem that roles were very permeable. At times, it was even the mechanics themselves who performed observations, as was the case with Luis Grosch during the 1865 solar eclipse, confirmed by the German journal *Astronomische Nachrichten*.[4] We also know that collaborators were often made to repair or adjust instruments when the mechanic was not present. Yet there's a very clear line between collaborator, mechanic and astronomer: while the mechanic performs the work that allows instruments to function (as we have seen in previous chapters), collaborators are those who ensure that these objects deliver the expected results. The work of collaborators therefore revolves around observations, measurements, photographs and recording and processing data. The functioning of an observatory requires the coordination and joint efforts of both groups, along with that of the astronomers themselves, who may sometimes perform calculations and

---

[2] Letter from Richard Prager to Enrique Matta Vial. Archivo Nacional de Chile, Fondo Ministerio de Instrucción Pública, vol. 3105, pp. 203–204.

[3] Archivo Nacional de Chile, Fondo Ministerio de Instrucción Pública, vol. 3105, p. 204.

[4] Moesta (1865, 300).

adjust, clean or even repair lenses, but whose work is focused on designing scientific problems that mobilize technicians and collaborators around research programs. In turn, there is a hierarchy between astronomers (thus the labels *first* and *second*), but a promise is also made to collaborators that they can take on future work that would make them worthy of being given the title of astronomers. Mechanics, with only very rare exceptions, continued to be technicians and could not have a professional career within an observatory.[5]

As Prager saw it, the labor conditions of collaborators and observatory staff—that is, their contractual status—defined the role they would play within the institution over time. In this sense, labor dynamics can provide clues regarding so-called "scientific practices" to the extent that we can glimpse what each person could or couldn't do. In other words: astronomy isn't a title or a university degree, but a labor practice.

Here we have selected three perspectives that allow us to better understand the work of collaborators: first, through contractually-defined limitations and the observatory's internal regulations, which reveal the predefined tasks for each rung in the hierarchy; second, through the destabilization of these limitations; lastly, through the silent place occupied by women's work.

The shadowy outline of collaboration work can be glimpsed today through testimonies on the conflicts that affected the stability of the observatory. Astronomers placed great emphasis on controlling their staff and avoiding errors in observations and calculations. The *day laborers of science* become visible to the extent that these efforts to domesticate the work of assistants, auxiliaries and computers did not bear fruit. Here, collaborators become something more than background characters. As we shall see, the astronomical enterprise was doomed to fail without them.

6.1 Disciplined Collaboration: Contractual Limitations

The first astronomical expedition by James M. Gilliss to Chile in the mid-nineteenth century, while primarily involving the US expeditionaries themselves, nevertheless required local support staff. The government named two students from the University of Chile, Ignacio Valdivia and Gabriel Izquierdo, to assist in "the operation of the instruments brought by the members of the expedition."[6]

In his memoirs of his voyage to Chile, Gilliss acknowledged the impact that these two Chilean students had on his work, as the US Astronomical Expedition "had only half the requisite number of assistants for an undertaking so laborious."[7] While the US scientists performed the astronomical tasks themselves, "there were several occasions on which one of the expeditionaries fell ill" and the Chilean collaborators "assisted in

---

[5] Regarding the hierarchies and division of astronomical labor, as well as the role of the material conditions of existence, see: Schaffer (1988).

[6] Keenan et al. (1985, 103).

[7] Gilliss (1855–1856, 507).

conducting observations."[8] When the National Astronomical Observatory of Chile was created in 1852 following the departure of the Gilliss expedition, it was resolved that these students would be the first collaborators for the new institution.

The founding of the Chilean observatory involved defining who would be in charge of observation and measurement work in support of its astronomers. What qualities and characteristics should they possess? On October 20, 1852, Ignacio Domeyko, the dean of the Physical and Mathematical Sciences Department at the University of Chile, sent a message to the Minister of Justice and Public Instruction in which he convinced him to hire an assistant and an auxiliary for the observatory. The requirements for these jobs were "that those considered for this position be in good health [and] have no other occupation."

From the beginning, the tasks of these collaborators were defined as follows:

1. Every day outside of their class time, as arranged by the two young men with the observatory director, they will present themselves at the office in order to carry out calculations based on the observations made, receive opportune instructions from the director and be available in case observations are to be done that night and to be responsive to the nature of these observations.
2. The assistant, in particular, must follow the uninterrupted course of observations and maintain an organized diary of this work, conducting them with the instrument that has initially been made available to the auxiliaries, and then with additional instruments when the director believes it apt for him to handle instruments of greater precision.
3. The assistant and the auxiliary will take note of the practical lessons given to them by the director and will procure to acquire the aptitudes needed to assist the director in his astronomical duties.
4. The director, for his part, will train them in the use of instruments at the earliest opportunity so that they can continue to make observations in the event that he falls ill.
5. Only in the event of illness may the assistant and the auxiliary refuse to assist in astronomical observations when so called by the director.[9]

There was already a hierarchy between auxiliary and assistant. While the former concentrated on performing calculations, the latter had to conduct observations and register them so that they could later be summarized. One can also note that they had no fixed hours, as their shifts would be decided by Director Karl Moesta. Beyond the question of having set hours, the value assigned by these regulations to the principle of "following the uninterrupted course of observations" is quite clear. Thus the importance of both the assistant and the auxiliary learning "the use of instruments" so that, in the event of an astronomer's illness, "observations can continue to be made."

Several days after he started work, as if echoing Prager's words cited at the beginning of this chapter, Ignacio Valdivia requested in writing that he "be permitted to

---

[8] Keenan et al. (1985, 103).

[9] Archivo Nacional de Chile, Fondo Ministerio de Justicia e Instrucción Pública, vol. 40.

absent himself twice a month, or perhaps not more than once" in order to continue exercising "the profession of surveyor when the occasion arises to conduct measurements in the field." Only under this condition would he "accept the position of assistant."[10] From the start, jobs at the observatory had to compete with other, better paid technical jobs. This situation made it practically impossible to respect the condition of exclusivity contained in the contracts. Not even Director Moesta himself could meet that condition. In a letter dated October 1852, Moesta informed the minister "that until the conclusion of the topographic work that has been assigned by governmental decree on August 26 of this year, I will not find the time necessary to give instructions to others, nor to drill them in handling the instruments."[11] The partial presence of the director doubtlessly made the training and supervision of collaborators even more complicated.

The observatory's dynamics made it increasingly necessary to precisely define not only the work of collaborators, but also that of the director. After over ten years of activity, during which the observatory was establishing a rhythm of constant work under Moesta's leadership, it became necessary to put the responsibilities for each role in writing. Moesta himself promoted regulations in 1865 that would lay out the responsibilities of the director and his assistants and auxiliaries.

It was established that the director must "contribute to the observatory's work and, to the extent that resources permit, to the advancement of astronomy and to the promotion and cultivation of this science in Chile," procuring to "provide those observations in which he takes part with the greatest possible perfection and exactitude." Furthermore, due to the observatory's position in global projects, the director was to "give preferential attention to those celestial phenomena that are only visible from the Southern Hemisphere, and those stars whose nature demands an observatory situated in the aforementioned hemisphere." The director must also "maintain relationships with certain northern observatories" and "send Chilean publications to other observatories and scientific bodies" as well as "ensure the conservation of the observatory's buildings, instruments and other tools." Finally, the director was required to "present a detailed annual report regarding the work executed over the previous year to the Ministry of Public Instruction."[12]

Collaborators and assistants had the obligation to "perform observations at the observatory three nights per week" and "provide assistance at the office from eleven in the morning until four in the afternoon." Their work was closely supervised, as they were required "to strictly adhere to the execution of the work assigned through the director's instructions" and "maintain proper conduct and express zeal and austerity in the work assigned to them by the director." In order to allow for better control and to facilitate their work, it was requested that "one of them, at the director's discretion, reside at the observatory; the resident assistant will care for the establishment's clocks and chronometers." Neither of the assistants "could be absent from the observatory for more than one month over the course of the year without license from the director,

---

[10] Archivo Nacional de Chile, Fondo Ministerio de Justicia e Instrucción Pública, vol. 40.

[11] Archivo Nacional de Chile, Fondo Ministerio de Justicia e Instrucción Pública, vol. 40.

[12] Archivo Nacional de Chile, Fondo Ministerio de Justicia e Instrucción Pública, vol. 148.

which may only be granted in cases of illness or other equally serious cases that the
director considers to constitute justified reasons." Finally, the regulations established
sanctions if they were breached: "In the event of improper conduct, negligence or
carelessness in the performance of their duties, absence without the permission of
the institution, insubordination or lack of respect toward the director, assistants will
be separated from their duties at the observatory by the Supreme Government upon
the director's request."

Article 11 required the director to "give a course on spherical astronomy and differential and integral calculus at the collegiate division of the National Institute." This
course, which was aimed at educating the institution's collaborators, was complemented with a provision in the following article that obligated Director Moesta to
"give lessons on practical astronomy once per week to the two most distinguished
students from the course on spherical astronomy at the National Institute: these two
students will be designated by the Supreme Government, in accordance with the
suggestions of the director." The observatory's collaborators were then differentiated from these students, who "will not be obligated to take part in routine work at
the observatory, but must be trained in the use of different devices or instruments in
order to perform various classes of observations and calculations." That they were
considered to be students prevented them from using the meridian circle and the
first equatorial telescope, but they were given access to "the reflection instruments,
theodolites, the universal instrument, the transit instrument, the second equatorial
telescope, a pendulum and a chronometer."[13] These instruments were normally used
for topographic and geodesic measurements.

It seems that these conditions were not substantially modified during the early
periods of the observatory's history. At the beginning of the twentieth century, with
the institution's growth and revitalization following the arrival of Friedrich Ristenpart
in 1906, it became essential to review the labor conditions of its employees. At the
same time, as revealed by bureaucratic records and even certain newspaper articles,
a series of conflicts between astronomers and the staff made it indispensable to
modify the regulations that guided the institution's work. As we have seen, it was
under Friedrich Ristenpart (1906–1913) that difficulties with collaborators escaped
the bounds of the observatory itself to become an issue of public debate. At the same
time, Ristenpart's tragic death, the loss of support with the change in government
and especially the economic crisis that shook Chile with the coming of the First
World War reduced the institution to one primarily oriented around meteorological
activities. This shift considerably reduced the number of collaborators, from 30 under
Ristenpart to seven in 1917. In 1929, the observatory became part of the University
of Chile, which implied a new stage in its history and a consequent redefinition of
tasks.[14] Under this new administration, its regulations would have to be modified.

At first glance, the 1929 regulations reveal a more complex institution, organized
into departments, as established in Article 2:

The observatory will be divided into three sections:

---

[13] Archivo Nacional de Chile, Fondo Ministerio de Justicia e Instrucción Pública, vol. 148.

[14] Keenan et al. (1985, 145).

1. Meridian Service
2. Equatorial Department
3. Astrophysics and Astrophotography Department.[15]

The new regulations delimited the tasks to be performed through the functions of these departments. Here we can also see how salaries and responsibilities were defined. While they didn't specify tasks for particular departments, they did delimit their responsibilities in accordance with their salaries:

Art. 3: The salaries of the observatory's staff will be as follows:

Director, Department Head	27,000–
Secretary—Accountant and Librarian	9,000–
Two Astronomers, Department Heads	21,000 each
Three Departmental Astronomers	15,000 each
Four Computers and Assistant Astronomers	9,000 each
Five Auxiliary Computers	2,400 each

The difference in remuneration is particularly marked in the case of the computers, who earned less than the doorman, whose salary was $4,000, never mind the precision mechanic, who earned $15,000. Even the housekeeper earned more than the computers, at $5,400.[16] Did this depend on their expertise or the time dedicated by these collaborators when compared to other employees? How was this regulation established?

Let us examine their responsibilities. The observatory director, apart from his responsibilities teaching "courses at the School of Astronomy," "annually presenting a memoir on the progress and current state of the observatory to the university rector," organizing scientific work, and "favoring and encouraging the production of original work by the astronomers," was also required to discipline the institution's staff. This involved actions such as "supervising the establishment and maintaining order and discipline within its walls" or "distributing employees throughout their respective departments, establishing regular work hours and assigning extraordinary tasks to be performed at the institution." It's clear that special attention was paid to the work of collaborators. Was this because of their low pay and the possibility that they did not perform their work with sufficient seriousness? Despite their importance for ensuring the exactitude of the observatory's observations and calculations, this was considered to be a minor task that was paid accordingly, as Prager had noted 17 years beforehand. Didn't their work require greater expertise than the housekeeper or the doorman? Why wasn't the essential nature of their work reflected in their prestige and remuneration? What do the regulations have to say about their responsibilities? What were their hiring requirements?

Article 14 established that "computers and other observatory employees will be required to be present at the observatory's offices for seven hours each day and must

---

[15] Archivo Nacional de Chile, Fondo Ministerio de Justicia e Instrucción Pública, no. 5459.

[16] Archivo Nacional de Chile, Fondo Ministerio de Justicia e Instrucción Pública, no. 5459.

carry out the tasks indicated by the director or the corresponding department heads." Meanwhile, "auxiliary computers will not be obligated to work at the observatory's offices, but must deliver a quantity of work equivalent to two hours per day each week, as assigned by the observatory director." So, was this job half-time, or at least part-time?

In terms of their expertise, Article 16 stated that "to be appointed a staff computer or assistant astronomer at the observatory requires a high school diploma, as well as continuing on to the Mathematics and Physics courses at the Pedagogical Institute or the School of Engineering at the University of Chile, and the Rational Mechanics and Astronomy and Geodesics courses at the latter school." It was therefore necessary to have prior studies, the same courses that the director himself taught at the university. Yet it wasn't a requirement to have finished these courses, as it was in order to be named departmental astronomer or head astronomer: "The courses indicated in the previous article must be completed along with a satisfactory evaluation on their final exams, as well as continuing on to the courses at the School of Astronomy mentioned in Title VII of these regulations." Auxiliary computers, as established by the following article, "must be regular students at the Mathematics and Physics courses at the Pedagogical Institute or the School of Engineering of the University of Chile."

As we have seen, regulatory provisions were the tools used by astronomers to ensure that the measurements and observations made by collaborators were trustworthy. Is this not what we know as *discipline*? And yet their practice refused to fit into these bureaucratic molds. How destabilizing were the actions of these collaborators?

6.2   Destabilizing Collaboration: Limitations Revealed

On April 26, 1853, less than one year after the National Astronomical Observatory of Chile was inaugurated, the first problems were already beginning to emerge. Ignacio Domeyko reported the complaints of Director Moesta about the poor condition of the establishment "due to the disinterest of its auxiliaries" to the university delegation. As Domeyko concluded, "I believe that, in reality, the problem indicated by Mr. Carlos Moesta is imminent and measures must be taken to radically remedy it."[17]

The complaint made by Moesta the previous month had made Domeyko see how the conduct of the assistant Ignacio Valdivia and the auxiliary Paulino del Barrio interrupted observation work:

> First of all, the assistant Mr. Ignacio Valdivia has been absent from the office not only in the fourteen days in December when there were exams at the National Institute, but also from January 14 to February 3, and then again on February 5, 8 and 9. Since February 9, he has attended somewhat regularly for two hours per day. In terms of the observations that I have assigned to this gentleman, he only performed a few of them up until February 12 and none since, refusing from the start to perform any observations that would occur after 11 at night,

---

[17] Archivo Nacional de Chile, Fondo Ministerio de Justicia e Instrucción Pública, vol. 40.

due to the late hour. Since February 12, however, he has refused to perform any observations at all, alleging that the wind has cut the power. The observatory auxiliary Paulino del Barrio has been absent from the office on February 5, 12, 14 and 15 and March 1, always due to illness. Calculations are performed in a most inexact fashion and it is a common occurrence that the simplest calculations have to be repeatedly sent back in order to correct serious errors.[18]

On the one hand, the observatory's normally nocturnal hours were not fully kept due to the conditions at the institution, which lacked a stable power source. On the other, exclusivity was a contractual condition that was only a reality on paper. As Prager diagnosed at the beginning of the twentieth century, the social situation—such as being a student or practicing other trades on the side—made absenteeism the rule rather than the exception. Along with potential illnesses that could affect the institution's limited staff, this meant that hours could go by without any observations being made. The end result was that calculations were inexact and needed to be repeated to ensure that they had been done correctly.

The greatest problem lay in the consistency of collaborative work, what Moesta himself referred to as "the steady and uniform progress of work" (underlined in the original). "Astronomical work, in itself, requires continual diligence and care, as the exactitude and value of observations depend on the pace of the chronometer and the position of the instruments," the director continued in his complaint to the university delegation. Work schedules would be worth nothing if these observation protocols weren't followed.

The size of the observatory staff and its willingness to work were key. Moesta informed Domeyko how, "in observatories in the other hemisphere," for every instrument there were "two assistants who are entrusted with the greatest responsibility for being present during certain hours of both the day and the night." Without this level of organization, the astronomer, argued, "the observatory will be considered to be a pastime rather than a scientific establishment." To work at an astronomical observatory, it therefore wasn't enough to have a background in mathematics or physics: one must also provide proof "of his perseverance, abnegation and love of work." Moesta was concerned about these setbacks as "a portion of the observations made over the course of the summer had to be summarized for publication."[19] How could the causes of these problems be explained? Like other astronomers and directors, Moesta did so through the disinclination of his staff for hard work—on the plane of morality, in other words. Just as the historiography has tended to overlook the role of labor conditions when explaining the development of astronomy, Moesta individualized the reasons his collaborators acted the way they did without considering the social situation in which they found themselves.

There's no question that the observatory's collaborators took sick days, just like any other worker, and this is made clear in the institution's bureaucratic documentation. In the 1857 annual report sent by Moesta to the Minister of Justice and Public

---

[18] Archivo Nacional de Chile, Fondo Ministerio de Justicia e Instrucción Pública, vol. 40.

[19] Archivo Nacional de Chile, Fondo Ministerio de Justicia e Instrucción Pública, vol. 40.

Instruction, he mentioned the irregularity with which collaborators came to work due to illness:

> Of the two young men who have been attached to the observatory, Mr. Gabino Vieyes has been unable to attend regularly, primarily due to an illness that ultimately obligated him to leave the capital; he currently serves as a mathematics professor at the institute in La Serena.[20]

On other occasions, however, the work of collaborators and auxiliaries was more successful, as the director emphasized:

> To date, Mr. Adolfo Formas has continued his studies at the observatory with an assiduousness and perseverance that deserves all praise. Despite his sickly physique, he has come to the office to the extent that he is able and performed his work in a satisfactory manner. Allow me to therefore particularly recommend him to you for your consideration, Mr. Minister.

One way of ensuring that collaborators and auxiliaries possessed knowledge of the observatory's devices and measurement procedures involved importing them from abroad. As with astronomers and technicians, the Chilean government reached out to Europe to acquire the expertise the country required:

> The Supreme Government, having deemed it wise to provide this establishment with an auxiliary who would particularly occupy himself with office work in order to provide greater celerity to the observations made at this observatory, a Mr. Arminio Volckmann has been hired by for this end by the Consul General of the Republic in Hamburg. Having been previously occupied as a computer for the astronomical almanac published annually by the Berlin Observatory, this new assistant has provided his services to the observatory office ever since his arrival in Santiago, that is, since this past August 22, and I am pleased to inform you that he performed his obligations in a satisfactory fashion up through this past January.[21]

Later on, however, the director reported his disappointment when, "for causes foreign to the observatory," Volckmann had begun to attend work "increasingly infrequently, entirely absenting himself from the establishment since early April, thus frustrating the higher aims that the Supreme Government sought to achieve."[22] Here the director does not morally reproach his countryman for his behavior, but simply expresses his disappointment that he had been unable to make use of his specialized labor.

Though the "causes foreign to the observatory" go unmentioned in Moesta's documentation, they correspond to Volckmann's side job as a surveyor, in which he formed part of the delegation that was registering the country's coordinates under the leadership of Amadeo Pissis. This situation evidently affected his work at the institution.[23]

In August 1858, several months after Volckmann's arrival in Chile, Director Moesta complained about his attendance issues, which had become a problem for the observatory:

---

[20] Archivo Nacional de Chile, Fondo Ministerio de Justicia e Instrucción Pública, vol. 84.

[21] Archivo Nacional de Chile, Fondo Ministerio de Justicia e Instrucción Pública, vol. 84.

[22] Archivo Nacional de Chile, Fondo Ministerio de Justicia e Instrucción Pública, vol. 84.

[23] Rosenblitt and Sanhueza (2010, XXI).

The work assigned to the aforementioned assistant since that date has been performed in a very imperfect fashion, as he has proceeded to handle the instruments with the greatest lack of care, unconcerned with the proper conservation of the books and making use of any pretext to excuse his failings.[24]

But the problem was aggravated further, Moesta argued, by Volckmann's even "more reprehensible" conduct:

He ignored the instructions I had given regarding the method for performing calculations and thus failed to comply with one of his primary obligations, as stipulated in his contract. Repeated reprimands calling on him to kindly perform his duty have been fruitless and when I had to chastise him again on the eleventh, he responded with such indecorous language that I saw the need to ask him to leave the observatory office in order to maintain order within the establishment.[25]

This wasn't just a matter of poorly performed work that could affect measurements: the greatest problem lay in his insubordination. This case reveals that, beyond any specific or personal situation that could be categorized as a labor issue, it is precisely hierarchy itself, targeted and disciplined, that is the very heart of astronomical activity. This contractual situation is what ultimately ensured the stability of the observatory's instruments and, therefore, that of the observations and data they produced. Thus the importance of the observatory functioning within the limits of the authority of its astronomers. To them, losing their authority would mean the end of their career. Unlike prestige, which some have argued represents a scientist's capital, an astronomer was validated before their collaborators by their hierarchical status.

It's not surprising to observe that the lack of assistants was the cause of the delays in the astronomical work assigned to the National Astronomical Observatory of Chile and that not infrequently affected the completion of the global projects in which the observatory participated. In the 1865 annual report sent to the Minister of Public Instruction, Director José Ignacio Vergara addressed the issue of collaboration as being key to evaluating the institution's performance. According to Vergara, in the preceding year "astronomical observation work had been interrupted for some time due to the lack of a primary assistant" and that it has only "been conducted regularly since this past January, when the individual who currently serves in this role was hired." Later on, he reported that "office work has continued without interruption, even though somewhat slowly as the observatory only had one assistant for most of the time that has gone by since April of last year." The report concluded by making a stipulation: "Likewise, if Your Excellency does not fill the position that is currently vacant, we cannot be expected to make much progress this year."[26]

The very fate of the institution depended on its staff. In his 1872 annual report, José Ignacio Vergara emphasized that:

The ordinary work at an astronomical observatory, particularly an observatory like ours that possesses such scarce staff, is of such nature that it cannot undergo considerable changes

---

[24] Archivo Nacional de Chile, Fondo Ministerio de Justicia e Instrucción Pública, vol. 84.

[25] Archivo Nacional de Chile, Fondo Ministerio de Justicia e Instrucción Pública, vol. 84.

[26] Archivo Nacional de Chile, Fondo Ministerio de Justicia e Instrucción Pública, vol. 148.

in the short interval of a year. As you know, the simplest astronomical problems that these establishments dedicate themselves to resolving require the involvement of many individuals over many years, sometimes centuries, of continuous observations.[27]

It was becoming difficult for the Chilean observatory to carry out the tasks it had taken on, continued Vergara, due to "insufficient staff," which had not "permitted, until now, the incorporation of nebula observations into the program, which are, as I have already mentioned, of absolute necessity for solving the great problems of the sidereal world."[28] How could global renown, or even a contribution to science, be expected of this institution when it lacked sufficient staff?

The labor conditions of collaborators continued to be a stumbling block, but one that was difficult to address. Two years later, Director José Ignacio Vergara once again mentioned the problem to the Minister of Public Instruction when discussing the need to build residences for the observatory's workers. These buildings, Vergara said, "while saving time and effort for observers, will likewise provide a guarantee of greater exactitude." At the same time, "due to the site where the observatory is located, it would improve conditions for employees and allow them to dedicate themselves to their astronomical studies with greater determination and austerity."[29] This was clearly the first attempt at addressing these problems that were holding the institution back. In this same letter, Vergara reminded the minister how, in the 1870 annual report, he had enumerated a series of factors that included but were not limited to the issue of housing:

> Another thing, Mr. Minister, that I must once again bring to your attention, due to its effects on the observatory's service, is the extremely disadvantageous situation in which our assistants find themselves, both due to the inferiority of their salary concerning their work and the special knowledge that is demanded of them, as well as due to the eccentric location occupied by the establishment, which lacks the lodgings required.[30]

Salary, type of expertise and proximity to the observatory (given that observations were normally performed at night) formed part of the social situation that had to be addressed in order to ensure the continuity of astronomical work. In this sense, the permanence of its collaborators was crucial. For Vergara, for "the work of a scientific institute, especially that of an astronomical observatory, to reach its full development and perfection" it was essential "to ensure, to the extent possible, the permanence of its employees, as it is not easy and for us it has been quite difficult to find individuals who are suitably prepared, so that variations in staff do not produce serious setbacks for the progress of our work."[31] The director then addressed the crux of the problem: "Regarding the assistants at the observatory, this situation has been completely neglected, which explains the frequent and nearly always detrimental changes in its personnel."[32]

---

[27] Archivo Nacional de Chile, Fondo Ministerio de Justicia e Instrucción Pública, vol. 148.

[28] Archivo Nacional de Chile, Fondo Ministerio de Justicia e Instrucción Pública, vol. 148.

[29] Archivo Nacional de Chile, Fondo Ministerio de Justicia e Instrucción Pública, vol. 148.

[30] Archivo Nacional de Chile, Fondo Ministerio de Justicia e Instrucción Pública, vol. 148.

[31] Archivo Nacional de Chile, Fondo Ministerio de Justicia e Instrucción Pública, vol. 148.

[32] Archivo Nacional de Chile, Fondo Ministerio de Justicia e Instrucción Pública, vol. 148.

Explaining this problem to the minister revealed the true responsibilities at the observatory beyond what was stipulated in the aforementioned regulations, not only in terms of quantity but also in kind and delivery times:

> According to our regulations, the three assistants must come to the office on a daily basis and occupy themselves with the calculations assigned to them for five hours: from eleven in the morning until four in the afternoon. The first assistant must also conduct astronomical observations at night for a period of no less than four hours, take part in meteorological observations and take responsibility for the establishment's library, as well as the adjustment and conservation of the pendulums and the chronometer. The second assistant must conduct magnetic observations at different times of the day and night and come in one or two nights a week, at the director's discretion, to conduct astronomical exercises, while the third must participate in making meteorological observations with the director and the first assistant and conduct practical exercises on the use of instruments with the second assistant.[33]

This doesn't just address those tasks involving observations and calculations, but also the conservation and adjustment of precision instruments (that is: their maintenance), as well as ongoing training, precisely in order to properly calibrate instruments, correctly make observations with telescopes and learn basic mathematics for performing calculations. Calculations, the product astronomers used to produce their articles and talks, were always weak if these obligations weren't adequately performed, in accordance with the protocols that the astronomers had taught to their collaborators. This is why Vergara requested something so basic as having rooms available for his staff:

> Dispensing with our frequent extraordinary activities, from the enumeration made above we can state that the time required for the everyday, ordinary work that is demanded of the observatory's assistants is: at least eight hours for the first and from six and a half to seven for the second and third. It must be taken into account that, as this time cannot be continuous, and as there are no lodgings for these employees at the establishment, they must make three or more trips per day over more or less long distances due to the eccentric location it occupies.[34]

What had begun as a request to build lodgings for the observatory's collaborators leads us to the issue of their labor conditions. According to Vergara, the current "state of things" meant that collaborators would arrive "always shaken and often exhausted at the start of their shift." This allowed him to easily understand that "given this attitude, aside from a multiplication of the personal errors that are inevitable in observations, they were not performed with calm and good will (...) on the part of the observer." For work to be performed optimally and especially so as to not destabilize observations, Vergara concluded that "this can only be achieved by offering them some personal comforts." Director Vergara had understood that the observatory was ultimately nothing but a workplace.

It's interesting to note the way perceptions of collaborative work influenced labor conditions and thus the treatment of staff. In 1881, José Ignacio Vergara wrote a letter to the Minister of Public Instruction in which he complained (again) about the ineffectiveness of an assistant. Yet this assistant was unlike the engineering students

---

[33] Archivo Nacional de Chile, Fondo Ministerio de Justicia e Instrucción Pública, vol. 148.

[34] Archivo Nacional de Chile, Fondo Ministerio de Justicia e Instrucción Pública, vol. 148.

that generally came and went at the observatory: Second Assistant Enrique Guzmán was none other than the deputy rector of the University of Chile. The problem was that he hadn't performed his afternoon observations. "Mr. Guzmán observed that he couldn't come in the afternoons because he had to focus on his labors as deputy rector at the university," the director reported. Vergara responded by citing the observatory regulations that clearly defined the shifts of its collaborators. Guzmán argued that he hadn't read the regulations, adding that "he had accepted the job because he had some hours free in the morning and wanted to take advantage of them to obtain some extra income, but from listening to what I had to say, he realized he couldn't serve the university and the observatory at the same time and would soon resolve the matter." Vergara continued to insist by writing to Guzmán and the minister, calling the former in to meetings at his office. Guzmán never kept the appointments, responding inconclusively in an attempt to delay:

> The great workload I have at the university prevents me from properly resolving what I must do in order to duly address my new responsibilities with which you have entrusted me. These considerations, combined with my right as a public servant to have one month per year for my own amusement, which I did not utilize last year, much less in this one, have driven me to delay the most suitable resolution I must make on this matter.[35]

After exchanging multiple letters with the minister and his second assistant, Vergara decided to bring the matter to a conclusion:

> Yesterday and today I have fruitlessly waited for you, and as I need to speak with you regarding your service at the observatory and your character as an employee of this establishment, as described in my previous letter, please inform me if you are willing to come or not; in the former case I will wait for you tomorrow until eleven in the morning and, in the latter, you will be forcing me to proceed in a fashion that would be very unfortunate and disagreeable to me.[36]

Evidently, proceeding in a "very unfortunate and disagreeable" fashion meant terminating his contract. Guzmán responded on December 19, 1881, not without expressing his anger:

> My dear sir:
>
> I have had the honor of receiving your letter, dated today, in which you have told me that you have fruitlessly waited for me and in which you indicate that you are willing to proceed against me in a very disagreeable fashion.
>
> The tone and language of your letter oblige me to inform you that I cannot give you an answer. While I may be an employee, I have the right to treatment that corresponds to my education.
>
> Yours truly,
>
> Enrique Guzmán.[37]

Was an employee really treated differently than an educated subject? Did one's educational level grant an institutional backing that permitted the mistreatment of

---

[35] Archivo Nacional de Chile, Fondo Ministerio de Justicia e Instrucción Pública, vol. 148.

[36] Archivo Nacional de Chile, Fondo Ministerio de Justicia e Instrucción Pública, vol. 148.

[37] Archivo Nacional de Chile, Fondo Ministerio de Justicia e Instrucción Pública, vol. 148.

those in a subaltern position? This case reveals the disrespect shown to those who occupied less prestigious roles at the observatory, which was in turn a reflection of the work assigned to them and its corresponding recognition.

6.3   Hidden Collaboration: Women's Work

One of the most fascinating chapters in the history of Chilean astronomy during the period under study, and yet one of the least well known, involves the role of women at the National Astronomical Observatory of Chile during the early decades of the twentieth century. Here, the question of work and the role assigned to an individual due to their gender becomes evident.

We are familiar with the cloak of silence that conceals the role of women, and not just in the history of science. As Michelle Perrot has argued, the difficulty of studying women's history lies above all in the fact that their tracks have been erased.[38] At the Chilean observatory, this was fully achieved through what some have called the masculine construction of scientific activity.[39] In the institution's history, the only episode in which women appear, as we shall see below, involves a female employee who had been treated unjustly by Director Friedrich Ristenpart.[40] Did nobody ask what work this woman performed? Does the historiography share the bias that women have nothing to do with *properly astronomical* work?

According to Perrot, given that the recording procedures upon which history depends are the fruit of a selection process that privileges what is public, women appear as an "intermittent light, rarely seen directly."[41] An early glimpse of women's work, then, can be found by reviewing the contracts and appointments of observatory staff. On March 16, 1911, Elisa Weber was named head photographer; Laura Neira, Luz A. Banda and Cristina Kröger as first photographers and Olga de Serre, Ester Maureira and Juana Giannini as second photographers. They were all assigned to the astrophotography department.[42] The following year, Zaira González was hired for the Meridian Service, Sofía Montiel for the astrophotography department and Teresa Herrera as second photographer.[43] In 1913, Dolores Fernández was taken on

---

[38] Perrot (1998, 43).

[39] Jones et al. (2022, 3).

[40] Keenan et al. (1985, 133).

[41] Perrot (1998, 48). It is essential to point out that despite the relevance and presence of women, as will be seen later in this chapter, their work is not mentioned in scientific articles produced by the Chilean Observatory nor in publications in Germany (such as *Astronomische Nachrichten*), nor Chile (such as *Anales de la Universidad de Chile*).

[42] Archivo Nacional de Chile, Fondo Ministerio de Justicia e Instrucción Pública, vol. 2762.

[43] Archivo Nacional de Chile, Fondo Ministerio de Justicia e Instrucción Pública, vol. 2953.

to measure photographic plates.[44] In 1916, the calculations department incorporated Sofia Montiel, Zaira González and Ester Maureira as second computers.[45]

Apart from these appointments, other administrative records in the archives shed light on women's work. On May 11, 1911, it was decreed that David Guarda, a computer at the astronomical observatory, be replaced by Ana Schubert.[46] Other replacements reveal that women also worked in calculations, as can be seen with Elena García, who was transferred from that department to astrophotography in order to replace Olga de Serre, who went on medical leave in October 1911.[47] There are also records of medical leave being granted to Teresa Herrera, whom we see working in the astrophotography department.[48] When she was transferred to the Meteorological Institute, we learn that Cristina Kröger had been working as the astronomical observatory's second draftswoman.[49]

All these decrees of appointments speak of names and contracts. Paraphrasing Perrot, it can be said that these records are but the visible flashes of women's work at the observatory. We not only don't know how they came to the institution nor what they did before they entered the world of astronomy, we're even ignorant of the very tasks they performed beyond the departments where they worked. Will we once again have to look for traces of their activities, as we have done with the mechanics, in the light cast by conflicts and public controversies?

In attempting to shed light on women's work in the early history of astronomy in Chile, we once again come up against the astronomer Friedrich Ristenpart and his notorious attitudes toward those under him. It all started with the complaint made by the mechanic Richard Wüst in December 1912 regarding the case of the selenium, leading to administrative proceedings against the director, which became a public scandal after it was covered by the press.

What was the basis of these proceedings and how did women's work make its appearance?

Wüst primarily complained about the irregularities in the receipts that were to be signed for unmade expenses that were charged to the observatory budget, as well as uncanceled per diems and travel expenses. Ristenpart defended himself by arguing that he had taken out loans in his own name to acquire scientific instruments and pay for the expenses of the new building at Lo Espejo. While he had been paid back on the principal out of the observatory's budget, he had incurred an interest that he had no reason to personally assume. All of these quantities, plus other income he had obtained from leasing out the observatory's buildings and carriages, had the objective of offsetting this loss.

---

[44] Archivo Nacional de Chile, Fondo Ministerio de Justicia e Instrucción Pública, vol. 3140.

[45] Archivo Nacional de Chile, Fondo Ministerio de Justicia e Instrucción Pública, vol. 3371.

[46] Archivo Nacional de Chile, Fondo Ministerio de Justicia e Instrucción Pública, vol. 2788.

[47] Archivo Nacional de Chile, Fondo Ministerio de Justicia e Instrucción Pública, vol. 2849.

[48] Archivo Nacional de Chile, Fondo Ministerio de Justicia e Instrucción Pública, vol. 2938.

[49] Archivo Nacional de Chile, Fondo Ministerio de Justicia e Instrucción Pública, vol. 3141.

These accusations, however, revealed other inconsistencies in the declarations of observatory functionaries. The administrative proceedings started off by interviewing the observatory's female employees, including Ana Schubert. She declared that she had not been paid when substituting her male colleagues in the calculations department, but that Ristenpart had nevertheless ordered her to sign blank receipts for payments that were never made.

The observatory's women, as these proceedings reveal, formed part of a strategy by Ristenpart to save money for the institution. The observatory librarian declared that he was aware that, in previous years, "Miss de Serres, Miss Maureira, Miss Neira, etc. have been paid salaries lower than they were owed, and that he is unaware of the destination of the discounted sums."[50] Later on, Ana Schubert declared that she received less than she was contractually owed. When asked why she didn't file a complaint, Schubert responded that she didn't initially know precisely how much she was owed and that the accountant had refused to respond to her queries. When Ana Schubert was asked if she was aware if things were different for the other women, she stated "that she has heard Miss de Serres, Miss Kröger and Miss Maureira say that they were being paid less than the salary they had a right to."[51] In turn, Luz Banda declared that, like Fernández, she had received less than she was contractually owed and that the difference was used to pay other women "who didn't have an official position."[52] In response to these accusations, "the director (…) said that these sums that had been discounted from their pay were used to pay other employees."[53]

The labor conditions of women at the observatory reveal that, to Ristenpart, their value as collaborators lay in their gender. While this condition limited their chances of advancing their scientific careers, it allowed the institution to profit from an educated workforce that was cheaper than their male counterparts.

Ristenpart defended himself from the accusation of paying male and female collaborators different salaries by stating that "all the ladies seemed satisfied" as "it was still more than what they were used to." Ristenpart even argued that "the ladies whose salary I reduced consented to this" (underlined in the original) "and as there were no complaints, being able to freely make use of their salaries (…) I believed I had committed no irregularities."[54] In other words: for the director, their lower salaries were a consequence of the belief that women deserved less pay for the same work.

Ristenpart argued that these women had received less training than their male peers. The case of Dolores Fernández is very illuminating here. During the administrative proceedings against Ristenpart, she declared that he had promised to transfer her from the photography department to the calculations department if she took algorithm classes. She stated that she had studied algorithms for several months with a private teacher, but that the director ultimately decided on a male computer.

[50] Archivo Nacional de Chile, Fondo Ministerio de Justicia e Instrucción Pública, vol. 3105.

[51] Archivo Nacional de Chile, Fondo Ministerio de Justicia e Instrucción Pública, vol. 3105.

[52] Archivo Nacional de Chile, Fondo Ministerio de Justicia e Instrucción Pública, vol. 3105.

[53] Archivo Nacional de Chile, Fondo Ministerio de Justicia e Instrucción Pública, vol. 3105.

[54] Archivo Nacional de Chile, Fondo Ministerio de Justicia e Instrucción Pública, vol. 3105.

Fernández declared that this wasn't required because "the computers who have the job I requested never learned logarithms and yet they always did their work exactly and correctly. Alberto Soza, who is in charge of assigning work to these employees, can confirm this."[55] Fernández had previously declared that she had substituted male colleagues in this department, even though she had been hired as a draftswoman. In his defense, Ristenpart reported that Fernández "begged me to give her a computing position that was about to become vacant, in spite of lacking the knowledge that this position requires." Ristenpart even declared that he had supported Fernández in spite of her incompetence:

> As with the other draftswomen, I lost a great deal of time myself in teaching her the most essential tasks for her job, without her ever managing to understand, and so the drawings she made on paper were nothing but utterly useless sketches. She also declared that she had no intention of learning this job, which was overly exhausting, and that she only wanted a position as a computer (as it has better remuneration). She often failed to show up for work at the office, whether with permission or without it, and yet I always had the greatest consideration for her as I understood that her health was very delicate.

Here we can see many familiar prejudices about women's work: it is believed that they don't understand the tasks assigned to them, that they have little interest in learning and that they are constantly stricken by health problems. Later on, Ristenpart summarized the problem by arguing that Fernández "begged (me) to give her the job she so badly wanted" and that he replied that "if she could prove to me that she possessed such knowledge, she would receive that job, even though she was the one who had been at the observatory for the shortest time (…) and lacked any degrees that would qualify her for this promotion."[56] While he demanded ability and knowledge of Fernández, the same proceedings reveal that male collaborators enjoyed a shorter workday "so that they can continue to study advanced mathematics." Nor were the astronomy classes that Ristenpart taught at the University of Chile, which were attended by practically all of the observatory's collaborators, accessible to women. The first Chilean woman to study engineering, Justicia Acuña, was just beginning her studies in 1913 and she would not receive her degree as Chile's first female engineer until 1919.[57]

It seems that Ristenpart believed that there were limits for women's work at an astronomical observatory. While women could play a role in the drawing and photography departments, they were restricted from the calculations department or from handling instruments directly. Nevertheless, this investigation reveals that women commonly worked performing calculations, as was the case with Elena García and Ana Schubert. Ristenpart had even included Elisa Weber, Teresa Herrera, Teresa Flores and Laura Neira as forming part of the calculations department in his 1909 annual report.[58] Calculations were likewise performed in the astrophotography department. In her deposition, Esthre Maureira stated: "I forgot to explain that in

---

[55] Archivo Nacional de Chile, Fondo Ministerio de Justicia e Instrucción Pública, vol. 3105.

[56] Archivo Nacional de Chile, Fondo Ministerio de Justicia e Instrucción Pública, vol. 3105.

[57] See Sanhueza-Cerda (2017, 27).

[58] Archivo Nacional de Chile, Fondo Ministerio de Justicia e Instrucción Pública, vol. 3105.

my position of stenographer I'm not limited to merely writing, as I am primarily occupied by the calculations presented by the measuring machines for photographic plates."[59]

Though not always successful, Ristenpart sought to limit the access of women to the observatory. In the conflict regarding Dolores Fernández's promotion, his reasoning evidently associated gender with role. Ristenpart argued that he had never imagined a job for her as a computer: "No, I promised Miss Fernández a promotion in her department (astrophotography) if she worked well. Promising her a job as a computer is impossible in itself, as computers work alongside the astronomers in the men's office, where there would be no place for a lady."[60] This is clearly contradicted by the fact that the annual report he himself had written several years beforehand included four women on the staff of the calculations department.

The director believed that a woman, no matter how highly qualified and hard-working, could not replace her male colleagues as astronomical work was seen as a masculine affair. Nevertheless, the reality of women's work at the observatory couldn't be further from this prejudice. If we study the data available on the volume of women's collective labor at the Chilean observatory in the early twentieth century, we come to realize that the initial seven women hired in 1911 were joined by another three women in 1912 and that the total number of women at the observatory had reached 12 in 1913, outnumbering men in the astrophotography department and equaling the number of men in the calculations department.[61]

At the same time, women's work at the observatory went beyond making photographs, drawings and calculations. Ristenpart's annual reports make it clear that women were involved in international projects such as the efforts to photograph Halley's comet in 1910 and the expansion of the Cape Photographic Durchmusterung (CPD), a global enterprise that began in the Northern Hemisphere in 1852. In terms of more local undertakings, the women of the observatory dedicated themselves to compiling stars for the Santiago Catalog and preparing outreach materials based on the observatory's research and astronomical expeditions, as well as on popular astronomical subjects. Yet outside of institutional reports, women at the Chilean observatory have never been credited as the authors of databases, observations or interpretive texts. In the 1910 annual report, it was mentioned that the observations of stars that would be used as the basis for the Santiago Catalog "had been compiled by Misses Weber and de Serres." The same report describes how "Miss Weber," using a photographic camera, "captured views of Halley's comet," of which "especially interesting is a plate taken on June 6 that reveals an eruption on the comet's tail in a retrograde movement. Dr. Zurhellen has sent the slides to Dr. Koff in Heidelberg."[62]

---

[59] Archivo Nacional de Chile, Fondo Ministerio de Justicia e Instrucción Pública, vol. 3105.

[60] Archivo Nacional de Chile, Fondo Ministerio de Justicia e Instrucción Pública, vol. 3105.

[61] It must be said that these figures do not include those women who lacked a formal contract, which could elevate the number of women still further.

[62] Informe Anual del Observatorio Astronómico Nacional 1909, Anales Universidad de Chile, p. 939.

To what extent was this collaborative work essential to the development of astronomical activities? Ristenpart's use of photographs and drawings representing his work outside of Chile, as well as at public talks; photographs and the calculations based on photographic plates; and, increasingly, the work of the calculations department make it clear that it was very important. In one sense, women did the work that men disdained. The case of Harvard Observatory reveals that one of the reasons that women worked in astrophotography was because one had to observe photographic plates for hours on end and so it was considered to be unspecialized office work that would bore the observatory's men. At the same time, Ristenpart saved money by having a group of collaborators that he could pay less than their male counterparts. Just as at Harvard, astronomy in the early twentieth century required many people and the observatory staff could not handle all this work, nor was there the budget to hire so many men, while the difference in salaries between men and women at the time made it more economical to hire women at scientific institutions. There were also many women during this period who were interested in science in general and astronomy in particular, which led many women to contribute to astronomical observatories voluntarily and *ad honorem*. Gender biases even assumed that the *feminine eye* would be better trained to differentiate between the brightness, tonalities and chromatic ranges of photographic plates, which explains the widespread acceptance of the idea that women should work with photographs in particular.[63]

The hidden nature of women's work at the Chilean observatory both during the period under analysis as well as in the subsequent historiography have mutually reinforced each other over the years. The history of astronomy in Chile has situated the emergence of women in the field among the so-called pioneers, that is, the female astronomers who established their careers in the mid-twentieth century.[64] Despite the public records found in the archives, technical collaboration has been overlooked. Put more directly: if it's already difficult to make women visible in science, it's even harder when we're discussing technical work.

In the end, the silence regarding women's work at the dawn of the century appears as the convergence of present and past traces that are erased by history, time and again.

References

Gilliss, J. M. 1855–1856. *The United States astronomical expedition to the Southern Hemispheres in 1849–'52*, vol. 2. Washington: Nicholson Printer.
Jones, C., A. E. Martin, and A. Wolf, eds. 2022. *The Palgrave handbook of women and sciences since 1660*. Cham: Palgrave, Macmillan.
Keenan, P.C., S. Pinto, and H. Alvarez. 1985. *El Observatorio Astronómico Nacional de Chile (1852–1965)*, 1985. Santiago: Universidad de Chile.
Moesta, K. 1865. Schreiben des Herrn Prof. Moesta, Dir. Der Sternwarte in Santiago de Chile an der Herausgeber. *Astronomische Nachrichten* 1555: 300.

---

[63] For further similar cases see Mullen (2020) and Stevenson (2014).

[64] Undurraga and Meier (2022).

Mullen, K. 2020. Temporary measures: Women computers at the royal observatory, Greenwich, 1890–1895. *Journal for the History of Astronomy* 51 (1): 88–121.

Perrot, M. 1998. *Les femmes ou les silences de l'histoire.* Paris: Flammarion.

Rosenblitt, J., and M. Sanhueza. 2010. *Cartografía histórica de Chile.* Santiago: Biblioteca Fundamentos de la Construcción de Chile.

Sanhueza-Cerda, C., et al. 2017. *100 años de la Escuela de Ingeniería y Ciencias.* Santiago: Universidad de Chile, Escuela de Ingeniería y Ciencias.

Schaffer, S. 1988. Astronomers mark time: Discipline and the personal equation. *Science in Context.* 2 (1): 115–145.

Stevenson, T. M. 2014. Making visible the first women in astronomy in Australia: The measurers and computers employed for the astrographic catalogue. *Publications of the Astronomical Society of Australia* 31: 1–10.

Undurraga, V., and S. Meier. 2022. *Pioneras. Mujeres que cambiaron la historia de la ciencia y el conocimiento en Chile.* Santiago: Centro de Investigaciones Diego Barros Arana.

Chapter 7
Conclusions: Traces of a Silence

Abstract The conclusions highlight how the history of astronomy in Chile has been shaped by silence—the absence of recognition for the technicians, mechanics, and workers who made astronomical research possible. These "day laborers of science" left only traces in archives, overshadowed by the astronomers celebrated as protagonists. Figures like Richard Wüst, a German mechanic at Chile's National Astronomical Observatory, remind us that scientific knowledge relies on mastering the material: observations depend on well-maintained instruments, precise calculations, and skilled hands. However, these contributions were omitted from historical narratives and buried in routine reports, contracts, and technical memos. Recovering these hidden figures is not just about justice; it reshapes our understanding of scientific work. Privileging the local context and labor structures reveals that astronomy was never the achievement of isolated geniuses but an entire network of workers. Studying their traces allows us to challenge the triumphalist vision of science and embrace a more complex human history.

Keywords Forgotten actors of history · Technical work · Chilean Astronomical Observatory

Technical work during the early period of astronomy in Chile has come down to us as traces of a silence, which we have sought behind the curtains of a history in which astronomers have been the protagonists. Is this another effect of that Greek inheritance that "has promoted, if not justified (…) the disdain for the artisanal and the *merely* technical in Western cultural history?"[1] In this tradition, manual labor is seen as a relationship of submission of man to matter that, as it subjects him to the designs of nature, prevents him from accessing the spiritual. The critique of Richard Wüst, the mechanic at the National Astronomical Observatory of Chile, that we should not forget that everyone at an observatory are but *day laborers of science* represents not just a revalorization of technical work in a society living in the wake of

---

[1] Kornwachs (2013, 30).

the Industrial Revolution, but also an assertation that the immaterial—knowledge—cannot arise without dominating the material. How can one obtain data and draw up tables if photographs aren't made correctly, if nobody reads them or if instruments are damaged by the elements? Wüst's appeal restores the place of objects in the generation of knowledge through praxis itself.

A question lingers regarding the expression *day laborer* used by Wüst, as it evokes one of the lowest categories of wage labor. Working for a day's pay implies a certain precarity and disrepute, but also a lack of education and formal training. How can we understand, then, that a mechanic trained at Germany's Zeiss workshops would feel so belittled in his working relationship with astronomers? Have we not seen precisely how important they were over the course of this book? The importance of the technicians, collaborators and employees who repaired and collaborated instruments, performed observations and calculations and built and maintained the observatory's infrastructure is undeniable. The *day laborers of science* fall under what Michel-Rolph Trouillot has called the four crucial moments in which silences enter the process of historical production: from the moment of "the making of sources," followed by "the making of archives," the retrieval of facts into narratives and, ultimately, "the moment of retrospective significance" or "the making of *history*."[2] These *day laborers* have not appeared in the papers of astronomers nor on their acknowledgements pages; they have not been recognized by the Chilean state, nor has there been institutional memory of their work, nor have they been recuperated by the historiography of the early period of astronomy in Chile.

It's worth asking who has taken the place of these *day laborers*. The answer is very simple: the astronomers themselves. The case of the female computers and workers in the astrophotography department in the early twentieth century, as described herein, is quite symptomatic. Despite their number and importance, they have not been taken into consideration when writing about the role of gender in Chilean astronomy. Was there no place for women in Chilean astronomy until the arrival of the first female astronomers in the mid-twentieth century?

Paraphrasing Zilsel, in what he has called the *metaphysics of talent and celebrity*, a social silence has surrounded the manual laborer in the history of Chilean astronomy. This metaphysics demands that other participants in scientific undertakings, such as technicians and collaborators, remain in the shadows, for only in this way can we isolate and understand the genius of the subject. Silencing other voices is part of this discourse. Through these reflections by Zilsel, we can understand Wüst's annoyance not just as a critique of the invisibilization of collaborative work at an observatory, but also as a way of making it clear that technical work is the basis of astronomical success.

Emphasizing the genius of the subject has allowed the historiography of astronomy in Chile, once again citing Zilsel, to raise up "individual figures of scientists as part of the discourse of the modern subject."[3] Only in this way can we understand the introduction of Chile, a peripherical country without a scientific tradition of its

---

[2] Trouillot (1995, 26).

[3] Zilsel (2008, 14).

own, into the history of global science (that is, Europe and the United States). So, if we interpret this case study through the presence of German astronomers and mechanics as the history of the European migrants who founded Chilean science, we would be repeating the history of science as a modernist feat. Evidently the United States and Germany (and France, too) are essential to understanding Chilean astronomy, but this book has shown that local institutions were part of the *locus* that these foreign scientists and mechanics occupied in the country. Local space isn't simply a passive place: for good or for ill (let us not forget Ristenpart's suicide) it has played a role in both channeling and hindering the work of astronomers, mechanics and collaborators. Likewise, it was the Chilean government that hired these foreign scientists and technicians. The Chilean state imposed a framework for the work of these Germans: it defined the relationship between astronomers and the University of Chile, decided where they would publish the results of their research and constantly evaluated their work. At the same time, the Chilean Congress oversaw the use of public funds and imposed political criteria on scientific work. On the other hand, scientific and technical work development was influenced by the emerging labor structures in Chile and the economic expansion driven by the mining and railroad industries. In this scenario, Observatory workers were not limited to operating scientific instruments but also had to adapt to a working system in which technical skills and practical experience acquired "on the job" were essential to their performance.

Privileging the local has been a way of uncovering traces of the silence surrounding technical work. These traces were always there in the archives, in documents recording the work of mechanics, collaborators and construction workers, but the exclusive interest in studying astronomers has led us to underestimate these documentary sources. Routine activities were captured in the observatory's annual memoirs: letters to the government minister who oversaw the institution, contracts, decrees and expense reports, as well as technical reports justifying the purchase of instruments or the hiring of new technical staff. This book has followed the everyday life of labor through these documents, but it has also studied moments of conflict, personal disputes, political pressure. By combining the ordinary and the extraordinary, we have been able to hear the voices of these *day laborers of science* underneath all this silence.

Understanding astronomy as labor allows us to have a more human vision of science. Here we have seen how the stabilization of observations and the processing of data were the result of the work of a heterogeneous set of people who depended on each other. Despite being deeply hierarchical institutions, no job is less important than another at an observatory. That we've only listened to the astronomers is a consequence of the silencing of some to exalt others. Here we've stopped seeing astronomy as the heroic effort of a few solitary individuals in order to better hear the murmur of its *day laborers*. Lorraine Daston has argued that the history of science, by distancing itself from the triumphal teleology that has dominated the discipline, has become an enormous disappointment to many scientists.[4] It's very probable that this book will be, too.

References

Kornwachs, K. 2013. *Philosophie der Technik. Eine Einführung.* München: Beck.
Trouillot, M. R. 1995. *Silencing the past: Power and the production of history.* Boston: Beacon Press.
Zilsel. 2008. *El genio. Génesis de un concepto.* Madrid: AEN.

---

[4] See: Does Science Need History? A Conversation with Lorraine Daston, Marginalia, October 28, 2022: https://themarginaliareview.com/daston-interview-p1/.

Author Index